Mankind and Deserts 2

Series Editor
Françoise Gaill

Mankind and Deserts 2

Water and Salts

Edited by

Fernand Joly
Guilhem Bourrié

iSTE

WILEY

First published 2021 in Great Britain and the United States by ISTE Ltd and John Wiley & Sons, Inc.

ISTE Ltd
27-37 St George's Road
London SW19 4EU
UK

www.iste.co.uk

John Wiley & Sons, Inc.
111 River Street
Hoboken, NJ 07030
USA

www.wiley.com

Library of Congress Control Number: 2020949240

British Library Cataloguing-in-Publication Data
A CIP record for this book is available from the British Library
ISBN 978-1-78630-631-9

Contents

Foreword

Mankind and deserts

Fernand Joly[1] departed from this world before he was able to complete this book, through which he had hoped to share his experiences of and passion for deserts.

"Yet another book on deserts!" some might think; another book to add to the numerous publications dedicated to these alien and fascinating worlds.

This book, however, is different from earlier books, as can be seen from its title "**Mankind and Deserts**". It is based on the singular relationships that are formed between humans and the world of the desert – relationships that are unique because they can be traced back to the very origins of humanity. Indeed, it is from the arid Horn of Africa (East Africa) that large migrations began and it is along the deserts, if not within the deserts themselves, that we find the major cradles of burgeoning historical civilizations. This inhospitable world is also associated with great spiritual leaders such as Moses, Jesus, Mohammed and the Buddha, while serving as the backdrop for adventurers and empire builders from Alexander the Great to Genghis Khan, or from the Incans to the Conquistadors in the Andes and Mexico.

What is this universe that is so barren and yet so mesmerizing?

"**All about a word**" was how Fernand Joly introduced his book: "**What is a desert?**" The ambiguity in this word results from the fact that it has been used in different senses across literature and throughout history. For a geographer-writer such as Fernand Joly, the one fact that stood out was that there was no one desert;

1 Professor at the University Paris VII, Denis Diderot, who spent 15 years at the Moroccan Institute of Science in Rabat.

instead there were multiple *deserts*, diverse and varied, ranging from Death Valley to the Kalahari, from the Namib to the Atacama or the Gobi desert. Each of these is a unique landscape, whose uniqueness was born out of its position with respect to the general atmospheric circulation, its geographic location with respect to the sea and to its relief features. And yet, transcending all differences, there is one constraint that binds them all together: **aridity**, defined as a natural physical state characterized by persistent dryness with the corollary of extremely scarce water resources. Both these concepts, **aridity** and **water**, are at the heart of the following chapters. Aridity (Chapter 3) because it "transcends time and takes over space" and water (Volume 2, Chapters 1–3) because it is the essential resource for all life, especially in the desert. Aridity is distinct from "drought", which is simply a "period with insufficient rainfall". Water is seen through the lens of how it appears on land: "wild water", which flows over slopes in an un-channeled manner (Volume 2, Chapter 1), under the impact of violent but brief downpours, and "concentrated waters", i.e. waters "concentrated" into a channel, fed by distant precipitation upstream of the borders of the desert. As can be seen, there is in fact a true **hydrography of the arid world**. Satellite images, among other sources, offer us clear and accurate reproductions of these systems: *fossil hydrographic networks*, the legacy of ancient humid periods, a map of intermittent water bodies (Volume 2, Chapter 2): *playas* and *sabkhas*, permanent lakes with fluctuating shorelines, such as Lake Chad or Lake Eyre, or large allogenous rivers (Volume 2, Chapter 3) that are born outside the desert but travel through the desert, sustaining life, such as the Colorado, the Niger and the Nile, "the first and most remarkable of rivers in the arid world".

The role played by salts in hot deserts is rarely discussed in a systematic manner. Guilhem Bourrié, geochemist and soil scientist at INRAE, has analyzed the origins and nature of these salts and demonstrated how important these salt deposits in the desert are for humans, whether they live off agriculture, livestock or, indeed, the salt trade (Volume 2, Chapter 4).

Chapter 1 of Volume 3, drafted by Joly, was edited after his demise by Yann Callot, a professor at the University Lyon 2 who is a specialist in ergs and dunes. This chapter examines the importance of **wind** in the desert. Wind, sometimes considered to be more emblematic of a desert than even dryness, counts among the earliest dynamics on Earth, an element that humans have not always been able to control. Indeed, this lack of understanding of wind has sometimes had disastrous consequences for certain projects (see the Green Dam in Algeria).

The final chapter in Volume 3, "**Living in the Desert**", was taken up by Marc Côte, Professor Emeritus at the University of Provence, who worked as a professor for 20 years at the University of Constantine. He has drawn on his deep knowledge of the land and the people of the Saharan region to present what he calls "The Desert Civilization".

Finally, it must be noted, with great regret, that, since 2010, "geopolitical turbulences have tended to change the fundamentals, to burn away knowledge and to prevent researchers from keeping in touch with this part of space and humanity."

Acknowledgments

Most of the illustrations were refined by Éliane Leterrier.

Yvette DEWOLF
Honorary Professor at the University Paris VII, Denis Diderot
Paris
October 2020

A man in the desert: Fernand JOLY Tademaït, Sahara, Algeria

Introduction: Water in Deserts

Because of the minuscule amounts of water and their meteoric variability, the discipline of arid zone hydrology is one of the highest forms of art and science.

V. Kotwicki, *Floods of Lake Eyre*

"Ce qui embellit le désert, dit le petit prince, c'est qu'il cache un puits quelque part..." (*"What makes the desert beautiful," said the Little Prince, "is that somewhere it hides a well..."*)

Antoine de Saint-Exupéry

It is because of its extreme scarcity, or indeed its absence, that water is so important in a desert, even when it is invisible, when one is searching for it or it is altogether lacking. It is exceptional to find open water sources (there are long periods when these open water bodies simply do not exist) and they are most often found at some depth. Finding water, collecting it, transporting it, using it carefully and storing it are all key issues to be resolved in desert life. Sometimes there is a lack of water because the excessive aridity results in deficient *hydrology* and a lack of *wild water* (Joly 2006), while at other times there is a poorly developed *hydraulic* system, such that domesticated water resources (Prosper-Laget 2001) cannot fulfill the needs of the population they serve.

Sometimes the areas involved are vast, like the Sahara, Arabia or central Australia; and, sometimes, there are smaller spaces where water, though present, is

Introduction written by Fernand JOLY.

unequally distributed, as is the case in central Asia, Mongolia, western North America and the Andes. The tragedy here is that across these regions there is a permanent imbalance between available water resources (highly limited) and the requirements of the ecosystem for ordinary consumption and development (Margat and Tiercelin 1998).

I.1. Hydrology – wild water

Hydrology is the science of water (Cosandey and Robinson 2000) and it is not a discipline that can be easily applied to deserts (Margat 1985). As with climatological data, this scientific field is hindered by a series of restrictions, arising, principally, from small human presence and the highly discontinuous nature of hydrological phenomena, which are scattered through time as well as space. There are, of course, a relatively large number of qualitative observations from nomads and travelers. However, these are often not very objective, difficult to access and unreliable. The quantitative observations that are archived are better monitored and valid, but they are far too scattered as they usually focus on inhabited sites: military or administrative stations, oases, mining or agricultural installations and construction sites.

Certain phenomena can also be studied using isotopic or magnetic labeling (Adar and Leibundgut 1995), through various mathematical or statistical models (Band 1985; Dassargues 2000) and, above all, through remote sensing (Mering 2008), especially phenomena such as floods, humidity in air and soil, and even the presence of subterranean water (Timmermans and Meijerink 1999). There have also been some systematic projects devoted to observation and measurement (Braquaval 1957; Dubreuil 1972; Roose 1977). However, apart from the large rivers, certain semi-permanent water courses and some large basins of economic interest (mainly related to oil), there has been rather a meager amount of data collected, and what does exist is fragmented, dispersed and patchy. This is partly because many desert zones lie in developing countries, which have economical constraints as well as a restricted qualified workforce. However, it is also because even among the large powers, such as the United States, Russia or China, equipment and regular tracking are allocated only to arid zones considered to be economically "useful" or profitable (Lacoste 2001).

I.2. References

Adar, E. M. and C. Leibundgut, eds. (1995). *Vienna Symposium, 1994, Applications of tracers in arid zone hydrology*. Red Books 232. Wallingford, UK: IAHS. 452 pp.

Band, L. (1985). "Simulation of slope development and the magnitude and frequency of overland flow erosion in an abandoned hydraulic gold mine." In. *Models in Geomorphology*. Ed. by M. Woldenberg. Winchester, Mass.: Allen and Unwin, pp. 191–211.

Braquaval, R. (1957). *Études d'écoulement en régime désertique - Massif de l'Ennedi et région nord de Mortchai*. Paris: ORSTOM. 92 pp.

Cosandey, C. and M. Robinson (2000). *Hydrologie continentale*. Paris: A. Colin. 360 pp.

Dassargues, A., ed. (2000). *Tracers and modelling in hydrogeology*. Red Books 262. Wallingford, UK: IAHS. 571 pp.

Dubreuil, P. (1972). *Recueil des données de base des bassins représentatifs et expérimentaux - Années 1951–1969*. Paris: ORSTOM. 916 pp.

Joly, F. (2006). "Les eaux sauvages des régions arides - Notions de base sur l'hydrologie des déserts." In *Geomorphologie: relief, processus, environnement* 4, pp. 285–298.

Lacoste, Y. (2001). "Géopolitique de l'eau." In *Hérodote* 102, pp. 3–18.

Margat, J. (1985). "Hydrologie et ressources en eau des zones arides." In *Bulletin de la Société Géologique de France*, pp. 1009–1020.

Margat, J. and J.-R. Tiercelin, eds. (1998). *L'eau en questions: enjeux du XXIe siècle*. Paris: Romillat. 301 pp.

Mering, C. (2008). "Analyse et cartographie des formations superficielles à partir d'images optiques et radar." In. *LES FORMATIONS SUPERFICIELLES, Genèse - Typologie — Classification — Paysages et environnement - Ressources et risques*. Ed. by Y. Dewolf and G. Bourrié. Paris: Ellipses. Chap. 11.2, pp. 413–439.

Prosper-Laget, V., ed. (2001). *Eaux sauvages, eaux domestiquées. Hommage à Lucette Davy*. Aix-en-Provence: Presses Universitaires de Provence. 341 pp.

Roose, É. (1977). *Érosion et ruissellement en Afrique de l'Ouest*. Travaux et Documents 78. Paris: Éditions de l'ORSTOM. 108 pp.

Timmermans, W. and A. Meijerink (1999). "Remotely sensed actual evapotranspiration: implications for groundwater management in Botswana." In *International Journal of Applied Earth Observation and Geoinformation* 01/1999 (3-4-1), pp. 222–233.

1

Water Falling onto Soil and the Effects It Produces

1.1. The arrival of water

We have seen the drastic conditions in deserts that govern the presence of water: an atmosphere deficient in humidity given the lack of water vapor and, consequently, the inability for air to reach saturation; scattered and irregular rains; and intense potential evaporation that is exacerbated by high daytime temperatures, annual temperatures and the ever-present wind. The very rhythm of rainfall in deserts is unfavorable to hydrology. Thus, the study of rainfall in deserts must focus on *how* the rain falls, as well as the effects that it has on the soil.

In deserts, fine and persistent rain, lasting several hours and impregnating the soil with moisture, is rare, although not entirely unknown. This kind of rainfall chiefly occurs in regions fed by humid air: sea breezes, monsoons and disruptions along the polar front. However, the more common type of rainfall in deserts is intense precipitation (sometimes in excess of 1 mm/min), concentrated into isolated showers (between 0 and 5, or 10 showers per rainy season), lasting a short time (between a few minutes and 1 or 2 h), provoked by rising convection currents, air rising over relief features, or by storms.

In Tamanrasset, for example, on a single day in September 1950, 44 mm of rain fell over 3 h (of which, 36 mm fell in 40 min), while the annual average for 15 years was 40.7 mm! These spectacular showers are, however, less violent than is sometimes assumed, and are certainly less violent, in most cases, than Mediterranean

---

Chapter written by Fernand JOLY.

showers. The most frequent showers over the Sahara bring in less than 4.5 mm over 24 h (Capot-Rey 1953), while at least 5–8 mm are required to effectively sustain vegetation, and an intensity of at least 0.5 mm/min is required to trigger off any streams following rainfall (Dubief 1959–1963). In any case, the maximum rainfall is, on average, less abundant in the middle of the deserts (between 13–25 mm/h) than along the semi-arid borders (between 35 and 50 mm/h) (Jansson 1982). Depending on the geographic location of these deserts, rainfall occurs either during one or two distinct seasons, or is distributed randomly over the year. Solid precipitation, in the form of snow, only plays a notable role in cold deserts or temperate deserts, or in high mountains in tropical regions. However, occult precipitation (mist, dew, frost) brings in a notable amount of water to the soil, occurring in almost all deserts at night, and through the day in fog deserts at high altitudes or along the coast.

Water arriving into the soil is an important step in the water cycle of a region or a specific site, from a hydrological and geodynamic point of view. In the desert, the effects produced by this water chiefly depend on the energy power of the rainfall, the angle of incidence of the rain striking the soil, the state of the green cover that comes in between the atmosphere and the soil, and the nature of the terrain. Green cover is largely scattered, producing a wide open space that is largely composed of bare rock. With very few exceptions, there is no grassy carpet and no trees. There are only bushes, low tufts or isolated individual plants. In these conditions, there is very little to absorb the energy of the rainfall or modify, or delay, its impact on the soil. Unless the wind blows in such a way that the rainfall falls very obliquely, the kinetic energy released by the shock of a raindrop falling onto the soil creates a *splash*, whose effects vary depending on the soil structure. On the hard surface of cohesive rocks, this splash only throws up water, producing a superficial and practically instantaneous flow with limited or no mechanical effects. On loose formations, however, the splash is softened but, through the impact, can bring about the destruction of aggregates, and the mobilization and saltation (see Volume 3, Chapter 1) of the constituent grains (at least those with a diameter smaller than 200 μm), which may be enough to cause *rain erosion* (Feodoroff 1965) in the form of the displacement of material over a very short distance. Even the gravel in regs may be destabilized and mobilized by the redistribution of the fine fraction that supports them. This results in flattening or residual compacting of the coarse elements and a spreading out of the fine particles; through glazing and the filling in of gaps this may lead to the surface becoming impermeable. At the same time, a part of the water that has fallen infiltrates through the grains and moistens the soil. The excess water runs off on the surface and/or flows into the subsurface (*subsurface runoff*, also called *hypodermic flow* (Tricart 1981)), into the interstices between rocks, following the topographic slopes. If the rain stops, the moisture front stabilizes over a few centimeters and then shrinks through evaporation. If the rain continues, surface runoff continues and the infiltrated water moves deeper.

A large part of this rain water returns to the atmosphere. Evaporation (see Volume 1, p. 113 *sqq.*) occurs immediately at the point where it fell and can even, sometimes, act upon the precipitation itself, preventing it from reaching the ground. The phenomenon is especially remarkable when the sun and wind follow a short but heavy shower. Further, the transpiration part of evapotranspiration is reduced to a minimum due to the relaxation in the green cover and, consequently, maximum water loss occurs through physical evaporation. Long intervals between brief showers, where the sky is clear or cloudless, give evaporation a clear field, because of the soil being directly heated by the sun's rays. This heating of the soil, however, is very uneven and depends chiefly on the soil's own properties, albedo, color, petrographic composition and the green cover it may have. A sharp difference in temperature is then produced between bare soil and the lowest layer of the atmosphere, which has a very low relative humidity (and, consequently, very high evaporating power). This difference is particularly striking in tropical deserts, being more variable in temperate deserts. However, even in temperate deserts the evaporating process is as effective, as can be seen in winter when there is noticeable sublimation of snow.

We cannot therefore deduce the hydrological consequences of water falling onto the soil by only studying the quantity of rainfall. Unfortunately, we lack measures that can quantify the combined effect of the complex factors that come into play from the time the water falls. However, there have been a few experiments with artificial rains over plots of land, which have provided some useful indicators on the subject (Yaïr and Lavee 1974; Roose 1977; Bryan and Luk 1981). Studying what happens to water after rainfall is only slightly easier. This analysis can help us establish the overall water balance and the circulation of water in the desert.

1.2. Weathering

Weathering is the series of mechanisms by which the atmosphere (air, water, various aerosols) act upon the rocks, modifying them and, eventually, breaking them up (Carroll 1970; Goudie 1989; Smith 2009). Weathering involves physical, chemical and biological processes. These processes operate together on the whole, however the extent of their individual participation differs based on local or momentary climatic, lithological and biotic conditions. We must also specify here how non-running water, brought in through rain, snow, mist or dew, plays a discreet but significant role in producing debris or waste, which are finally fixed or displaced by their own weight, streams and/or the wind.

1.2.1. *Thermoclasty*

Given the absence of water, at least for long periods of time, it has long been believed that erosion in deserts occurs only due to wind. However, the destructive force of the wind by itself is actually limited (see Volume 3, Chapter 1). Thus, many explorers and researchers surmise that in an arid environment, the direct action of variations in temperature may be one of the leading agents behind the fragmentation of rocks and this is expressed by the term *thermoclasty* (Walther 1893; Coutard 2008). However, there is a distinction made between *fissuration*, where cracks open up without the material dissociating, and *fragmentation*, where there are open fissures, and flakes or chips break away.

It is true that temperature differences in deserts are quite extreme due to the dryness of the atmosphere and the bare soil, which facilitates radiation. These temperature differences include seasonal differences in the high and middle latitudes, but also the sharp daily differences, especially in summer and in hot deserts. Around the middle of the day, soil temperatures can easily exceed 50°C or even 60 to 70°C (Peel 1974) (depending on the rocks, their own albedo, the biotic environment and exposure to the sun), while late-night temperatures often fall below 20, 10 or even 0°C. It is not uncommon to have a temperature difference spanning 30–50°C in a single day. The differences become much smaller as we go deeper into the soil, but can sometimes be noticed up to a depth or 30 or even 50 cm, especially in porous or fissured rocks into which air can penetrate. These temperature differences are destructive due to the frequency and intensity of diurnal expansion followed by nocturnal contraction, especially with respect to the tensions resulting from the rapid cooling following sunset. Cold rain falling on overheated soil acts in the same manner. The earlier disintegration of rocks also comes into play, in ways or timeframes that may not always be measurable; many fissurations are fossil fissurations, resulting from processes that are no longer present, such as tectonic movement or the freezing process.

When rocks have existing fissures, thermoclasty acts upon them through the alternating expansion and contraction of air trapped in the cracks. This varies depending on the nature and color (albedo) of the rocks. In compact rocks, the expansion and contraction of the rock itself, along with that of its mineral constituents, plays an essential role. In grained rocks (granite, sandstone) and crystalline rocks, the differential expansion of the grains or crystals triggers the granular disintegration that results in arenization (Godard and Bourrié 2008). On rocks that are covered by desert varnish or a crust, the different physical reactions of the rock and of its outer layer produce a scaling or exfoliation of the superficial layer. The case of loose gravel found on regs (Volume 3, Figure 1.1, p. 31) is a little more complex (Bertouille *et al.* 1979): small pebbles (1–4 cm) are likely to be most affected by fissuration (Soleilhavoup 1977), while larger pebbles and more voluminous stones (larger than 8–15 cm) are more likely to be affected by fragmentation (Joly 1962).

It is difficult to gauge the actual effect of thermoclasty in the field and its effects are rightfully debated. No one can deny that there exists a temperature differential between day and night, in deserts, that is steep enough to cause a thermal shock that can weaken the structure of a rock. This shock must be repeated often enough to produce an actual breaking up of the material. Experiments carried out in laboratories using different protocols and on various kinds of rocks (Coutard and Journaux 1976) have demonstrated that in dry conditions, with a temperature difference of the order of 50°C, at least 8000–20,000 successive shocks are required to visibly start off fissuration in grained rocks such as granite or sandstone, or in compact rocks such as basalt. Indeed, thermoclasty weakens the rock's resistance and exposes the rocky material to future fragmentation. But while its existence cannot be denied, its effects are limited to specific cases where there is a conducive convergence of microclimatic, lithological and biotic conditions, along with pre-existing fissuration. In any case, the fragmentation of rocks in deserts cannot only be attributed to this phenomenon.

1.2.2. *Hygroclasty*

The presence of water significantly increases the efficacy of fragmentation due to weathering. In this regard, the rain we consider is not so much the brutal showers, but the fine, "moistening" rains, poorly drained water and, above all, mists, dew and condensation of any kind. The factor that is most strongly implicated in the process is not variation in pluviometric volume, but rather the variation in moisture and the moistening or humectation of rocks, which can sometimes start off a chemical weathering (hydration) where some elements are dissolved. Due to this, the term *hygroclasty* was proposed for this particular process (Joly 1962). Studying this phenomenon is rather difficult, especially with respect to data collection and the fact that there are multiple factors, apart from the microclimate, that come into play. These include biological activity, the nature of the rock, its albedo, pre-existing fissuration in the rock, its porosity and the capillarity of the moistened zone (Smith 2009).

Fissures grow in size chiefly through mechanical means, which, like thermoclasty, consist of compressions and decompressions that affect the rock and the fluids within the voids. These mechanical means provoke a slow degradation through the disaggregation of grains and crystals in grained rocks; the scaling away of the walls of the cracks, joints and fractures in compact rocks (Freytet 2008b) results in an enlargement of the channels and greater circulation of air and water, which produces a clear fragmentation of the rocky material. In many cases, the role of past events, in more humid periods, has been preponderant. The phenomenon is simply continuing, albeit, more slowly, in present conditions.

As with thermoclasty, the effects of hygroclasty are difficult to evaluate (Peel *et al.* 1975), and laboratory experiments are still rare or too specific (Pissart and

Lautridou 1984) to allow for generalizations. Field observations can, nonetheless, provide a few useful indicators. Certain limestone pebbles, especially marlstones, which are either isolated on a reg or contained within a loose formation that can retain moisture for a long time, present polygonal fissuration, somewhat similar to "crocodile skin", or fragmentation resembling a "bread crust" (Joly 1962). The open cracks, which may be several centimeters deep, have sharp, clear-cut edges that are flared at their upper extremities; their walls are often flaking, marked by wavy lines (vermiculated), polished by the wind, or covered by a varnish. The intervention of water in these rocks can be clearly seen. They are morphologically analogous to the cracks and fissures caused by shrinking in the clayey or clayey-silty surfaces of playas and sebkhas as they undergo desiccation. The fragments that break away are themselves vermiculated, marked by raised, spalling domes, or glazed. Similar shapes and forms can also be seen on bare limestone slabs or crusts, which are highly exposed to humid winds. On surfaces that are cool or lie in the shade (which, consequently, retain moisture for a long time), scaling and granular disaggregation also cause residual cavities of varying depths, called *tafoni* (singular: *tafone*). The size of these *tafoni* can range from a few decimeters to a few meters, and they can lead to deep steep ridges, and even the formation of hollow rocks and arches of all sizes (e.g. in the deserts of the American West, or in the central Saharan region). Rocky walls and isolated rocks, which are chiefly attacked at their base, which are most exposed to excess moisture, often present basal grooves, and take the shape of sculpted rocks, or mushroom-shaped rocks. Despite being partly fossilized and incompletely formed, these shapes are still vivid and often truly spectacular.

1.2.3. *Cryoclasty*

Cryoclasty is the breaking or fragmentation of rocks caused by the freezing of water within gaps in the rock (Ozouf 2008). When discussing thermoclasty, we saw that the strain the rock undergoes during rapid cooling can lead to fissuration. In addition to this, when water present in the cracks and pores in the rocks freezes, it causes an increase in volume as the water changes from liquid state to solid state. This increase in volume can exert considerable pressure on the walls of these cracks during the freezing process, and the subsequent thawing results in a decompression that can cause the rock to shatter (Prick 1999). This "frost weathering" or "rock-splitting freezing" does not require very low temperatures and can occur even at temperatures of –2 or –3°C. The *gelifraction* process, however, which produces debris (*gelifracts*), requires multiple freezing/thawing cycles, oscillating around 0°C.

Cryoclasty chiefly affects rocks that are most sensitive to the effects of freezing, i.e. rocks in which freezing water is constricted by the narrowness of the voids. If the voids were not entirely filled, the freezing could occur easily without exerting pressure on the walls. In coherent rocks, cryoclasty operates through the discontinuities created by tectonic movements (cracks, faults), mineralogical assemblages, sedimentation (joints, stratification planes) or previous weathering, as well as the network of cavities (pores) and channels that open to the outside. In loose rocks, gelifraction primarily occurs at the interface between frozen soil (gelisol) and non-frozen or thawed soil (mollisol), and moves along the migration path of the frost front. This is complemented by vertical fissuration caused by shrinkage cracks that result from the drying that accompanies freezing.

In many hot deserts around the world, the *current* action of frost is quite limited. This is because the frost occurs for too short a period, or too rarely, to produce a significant number of the freeze/thaw cycles that are required for the mechanical disintegration of rocks. Further, the total volume of precipitation as well as the duration of the precipitation is generally too small to provide enough water for gelifraction or freezing. In cold, subpolar deserts, with long winters and permanently frozen soil (permafrost), there is no alternating cycle of freezing and thawing. This cycle is chiefly produced in zones where the gelisol and mollisol come into contact with each other, during autumn (freeze) and in spring (thaw). Paradoxically, this cycle is most active in temperate deserts, where winters are harsh, where the freeze/thaw cycles are more frequent, and where melting snow and atmospheric humidity contribute to the voids in rocks getting filled up. This can be seen notably in the deserts of Iran, in Central Asia, Mongolia and North America. The phenomenon can also episodically be seen in the septentrional Saharan region; in Negev; in the deserts of Syria, Iraq and northern Arabia; and in Pakistan, South Africa and southern Australia. In deserts that are continually hot, as in the central Saharan region, in Arabia and across most of Australia, cryoclasty is rarer, and is linked to cold climatic crises or associated with particular conditions, such as high altitude (Hoggar, Hedjaz, the northern Andes), lying in the shade and zones of maximum condensation. However, this activity is all relative. It is, naturally, more tangible along the subarid and semi-arid regions where the water supply is relatively more abundant. But humidification and gelifraction do not penetrate very deep in any region. Superficial cryoclastic formations found today are usually not very thick, measuring just a few decimeters at most. Indeed, we only largely find gelifracts left over from earlier ages, covering older and more abundant cryoclastic debris.

1.2.4. *Haloclasty*

Haloclasty is the disintegration of rocks caused by the action of evaporite salts, chiefly gypsum ($CaSO_4 \cdot 2H_2O$) and "salt" or halite (NaCl) (Evans 1970; Cooke 1981). It is known that deserts are rich in saline formations due to their geological past (Paleozoic, Triassic, Cenozoic) and from present-day deposits (coastal shores, salt lakes, playas, sebkhas, salt-affected soils) that are easily formed due to the evaporating climate. Salts and saline particles originating from volcanic activity, the sea and lakes, salt outcrops and deposits, and salt-affected soils, are transported by the wind, sea spray and rain. They are deposited by fog, dew and streams and are scattered across rocks and within their interstices. Fortuitously, in hot deserts as well as in cold deserts, the overall lack of water protects these formations from disappearing through being dissolved.

The disintegrating action of salts chiefly works in two ways: one of these is the crystallization of salts following the evaporation of a saline solution that had entered fissures or cavities in rocks. As with ice, the passage from an ionic state to a *crystalline* state results in a considerable increase in volume and may occur in any kind of rock. The pre-fissured pebbles on regs are, thus, broken into pieces by the distension of their existing microfractures or microjoints. The second way is through the *hydration* of saline minerals. The classic example of this is the transformation of anhydrite ($CaSO_4$) into gypsum ($CaSO_4 \cdot 2\,H_2O$), and this operation is also accompanied by a volumetric increase in the hydrated rock. In both cases (crystallization as well as hydration), the action that leads to the rock breaking up, like with freezing, is the dilation of an interstitial body that can exert pressure onto the walls of the gap. This pressure is comparable to that seen in cryoclasty (Prick 1999).

Various experimental research studies (Tricart 1960; Goudie *et al.* 1970) have shown that haloclasty is more effective than thermoclasty and is even more dependent on lithology than hygroclasty. However, it is a phenomenon that is widespread, due to the wind dispersion of salty dust.

1.2.5. *Biodegradation*

In the desert, all living things are heavily dependent on meteorological conditions, both for their vital development as well as for their distribution over and in the soil. Reciprocally, their presence and dynamism create important counterparts for all kinds of weather formations that are "always carried out in association with biological activities" (Ozouf 2008).

Neither plants nor animals can, by themselves, fissure or fragment hard, coherent, granular or compact rocks. However, they are able to occupy existing cracks and cavities in the rocks and can, thus, participate in the transformation of the rocks. *Bioclasty*, such as the fragmentation of a rocky outcrop through the action of vascular plant roots filling up a crevice, for instance, may be a spectacular occurrence, but is simply the result of each individual species making use of a given biotope, in a favorable microclimate, to optimize the space in which it has implanted itself. We can also cite here the example of isolated pebbles on an erg breaking up, through a direct shock or through ricochet, when mobilized during the passage of a group of people or animals. And, of course, there is the loosening and destructive effect of burrowing animals on loose rocks or, on the other hand, the consolidating and stabilizing effect of mound-building animals such as termites.

In reality, the action of living beings on the desert is much more complex and goes beyond the mechanical breakup of rocks. Microorganisms in particular, such as bacteria, fungi, algae and lichens, are found everywhere (Freytet 2008a). Their metabolism produces a large part of the active chemical elements found in the environment, such as O_2, CO_2 and various corrosive organic acids, which modify the pH of the milieu and control the unfolding of various processes (*bioalteration*). In this way, these organisms participate in the liberation of certain elements, in the formation of new mineral combinations (neo-formations), and the fixation of part of the products of alteration such as silica, oxides and hydroxides of iron, aluminium and manganese. In the mineral world, despite the harsh climate, living beings act as one of the principal factors in the destruction of rocks.

1.2.6. *Chemical changes*

Generally speaking, chemical alterations are governed by the presence of water and accelerated by high temperatures (Millot 1964; Muxart and Birot 1977; Duchaufour 2004; Bourrié, Trolard, and Freytet 2008). Some reactions are destructive through simple removal of matter, while others are more complex and involve the participation of living microorganisms, the formation of new minerals, the exportation of mobile elements or deposition of residual substances. In the desert, these chemical processes are more common than they would seem, given the aridity of the climate and the sparseness of vegetation.

The following principles underlie these reactions:

– We have already spoken of the *hydration* of rocks when discussing hygroclasty and the transformation of anhydrite to gypsum. This process involves the addition of water molecules to anhydrous or hygroscopic minerals, i.e. minerals capable of absorbing water, such as gypsum, marls or calcareous marls, or certain clayey formations that can "swell". This addition of water leads to an increase in volume and the loosening of the affected rock. Dehydration, on the other hand, results in contraction of the rock, with the material hardening and shrinkage cracks widening. Water can be brought in by rain but, most notably, comes from dew and morning mists.

Hydration is more effective at lower temperatures, i.e. at night and in winter (condensation), and dehydration occurs more commonly at high temperatures, i.e. in the daytime and in summer (evaporation). Arid climates, whether temperate or warm, are most conducive to the alternating "hydration/dehydration" cycle. Hydration involves the processes of hygroclasty and haloclasty in the disintegration of rocky outcrops, and the creation of certain minor desert forms, such as spalling, exfoliation and arches along the base of rocky structures or, again, the network of polygonal cracks along surfaces covered in fine-grained, dry substances.

– *Dissolution* is the homogeneous dispersion of a solid substance in a liquid (water). Dissolved substances are carried away as solutions or are absorbed by living beings. The most soluble substances (salt, gypsum, limestone) are carried away by infiltration or runoff. The least soluble material (dolomite, silica, clays) are left behind as residue (e.g. ruiniform relief) or combine to produce neoform minerals.

Solubility varies depending on the nature of the mineral, as well as on the temperature and pH of the water. Solubility is greater in carbonated rocks (limestone, dolomite) and in cold water (temperate deserts), due to high levels of dissolved CO_2, and is more diversified in siliceous rocks (igneous, metamorphic, vulcanites) because of their complex structures. These rocks dissolve more completely in tropical regions (hot deserts) as silica is more soluble at higher temperatures.

– *Hydrolysis* is the decomposition of minerals in a rock by water. Hydrolysis reactions depend, above all, on the quantity of water and on the temperature. In carbonate rocks, the most common occurrence is the formation of a solution. In siliceous rocks the basic chemical mechanism at play is the alteration of silicates. Hydrolysis varies depending on the degree of solubility of the rock components, especially that of silica (SiO_2), which increases with temperature and pH of the solution. Quartz (crystallized silica), which has very low solubility and also takes a very long time to dissolve, can last for a long period as relief features or as debris of different sizes (gravel, sand or silt). Quartz is most often the principal material making up the sand in large ergs. Carbonate rocks are eliminated through drainage; soluble silicates are carried away as solutions in groundwater and in streams. Insoluble substances and Fe and Al oxides form a residue and, when silica is not completely eliminated, this produces neoformed minerals: kaolinite ($2\,SiO_2 \cdot Al_2O_3 \cdot 2\,H_2O$), gibbsite ($Al(OH)_3$) or swelling clays (smectites, e.g. montmorillonites).

– *Oxidation* is either a process leading to the incorporation of oxygen or the liberation of electrons (e.g. $Fe^{2+} \longrightarrow Fe^{3+} + e$) at the level of the components of the rocks. This process particularly concerns ferromagnesian minerals found in igneous and metamorphic rocks. Oxygen, in the form O_2, is supplied in sufficient quantities, both in gaseous state (air, bioactivity) as well as in dissolved state (rain, mists and nocturnal dews). In both states, O_2 penetrates into the voids in the rocks and freely circulates (or through capillarity) in porous and aerated media like sand dunes or sometimes poorly cemented sandstone. The absence or scarcity of any organic matter (which would reduce available oxygen), high temperatures and alternating periods of humidity and dryness promote abundant and rapid crystallization of oxides, especially iron oxides, which lend a vivid color to the altered material: ochre (goethite, FeOOH) or red (hematite, Fe_2O_3). This is called "rubefaction" (not to be confused with color found in legacy sediments from older ages – lithochromy). Alkalines (Na, K) and earth-alkalines (Mg, Ca) and liberated silica are carried away by water, depending on their availability (Figure 4.2).

While chemical alterations are highly limited in all deserts because of the dry air and low humidity, there is greater activity (although slow and discontinuous) in hot deserts than in cold deserts. In present-day nature, they are almost entirely restricted to the subdesertic and subarid borders of hot deserts.

1.2.7. *Desert varnish*

Bare, exposed rock in the desert, either rocky walls or scattered rocks, are covered by a fine, dark mineral layer that gives the impression of being "polished" or "varnished". This phenomenon is widespread enough for people to talk about a "desert livery". The coating on limestone walls is ochre, red or bistre (brownish-yellow). Siliceous rocks have a darker shade, brown or black, with gleaming gray streaks. This "varnish" is not very thick, ranging from 2.5 µm to 2–4 mm, rarely thicker. It is largely (60–80%) composed of clay minerals (predominantly illite and montmorillonite), with the rest being formed of residues from oxides or hydroxides of iron or manganese and, depending on the underlying rock, silica or carbonates (Dorn and Oberlander 1981; Oberlander 1994; Freytet and Verrecchia 2008).

The formation of this varnish was initially attributed to purely physico-chemical processes, resulting from a sort of "sweating" or progressive exudation of the products of alteration from the underlying rock. When the rock was moistened by rain – or, more often, by dew – hydrolysis reactions were triggered, breaking up and destroying the mineral structure. Soluble elements, carried away freely or through capillarity, were eliminated or deposited on the surface through concentration (evaporation), oxidation (crystallization) or recombination (neoformation). High temperatures and scarce vegetation are principal factors in promoting the creation of solutions and the migration of silica which, along with manganese and iron, is the most commonly found element in the "desert varnish". This varnish is therefore made up of elements that are largely local to the deserts or have traveled very short distances, with some external input of material brought in through the wind, sea spray or rain. Despite this theory of chemical reactions, however, it was observed that there is no systematic continuity between the weathering front and the varnish and that, in many instances, varnish lies directly atop whole or unmodified rocks. As a result, researchers have long since (Laudermilk 1931) considered the role of living organisms in the process and, indeed, this is the current reigning theory on desert varnishes (Freytet and Verrecchia 2008). The organisms in question are microorganisms that form stromatolites, which would explain the layered structure that is characteristic of many varnishes (Dorn 1984).

Desert varnish can be found on all kinds of rocks, but is present particularly on crystalline siliceous, metamorphic or volcanic rocks, rich in a variety of elements. The varnish is thinner and less evenly distributed on alkaline rocks that are less diverse and more easily soluble. Porous rocks are more often varnished than compact rocks, while varnish on coarse-grained rocks, with large particles, (granites or certain types of sandstone) is very limited due to the granular disaggregation seen in these rocks.

Desert varnish can be found in many regions around the world, even in places that are not, strictly speaking, deserts. The phenomenon can be seen as far as altitudes exceeding 5,000 m in the Himalayas or the Andes, in tropical regions as well as in the Antarctic, and in regions where the temperature soars above 40°C, as well as in regions where it plummets below zero (Freytet and Verrecchia 2008). The formation of this varnish requires enough water to trigger processes of chemical weathering as well as the migration of elements. However, too much water will result in the total exportation of these elements. Thus, the process also requires dry periods that are long enough to facilitate the fixation of these elements on the surface of the rocks but not so long that the fixing organisms are killed. It appears that the best conditions for varnishing are those seen in regions with subdesert aridity and moderate desert conditions (both hot and cold) with a pluviometry that is less than 200 mm and a long dry season. However, we must also take into account legacy, in the case of varnishes produced in earlier times and under climatic conditions that were different from the present-day.

In the Sahara, it is has been often observed that the older the desert varnish, the thicker and darker it is. It has also been seen that the generalized varnish coating does not extend to outcrops that emerged more recently. Similarly, paleolithic tools scattered across regs carry this coating (at least on the exposed surface), while mesolithic and neolithic artifacts do not. Thus, the time elapsed since the appearance of the last varnish is likely to be around 10,000 years. However, despite the progress made (most notably in Californian deserts) using isotopic methods (Dorn *et al.* 1986), cases of precise dating are too few and far between to allow researchers to develop a general theory on the subject. What remains, therefore, is the possibility of using the varnish as a relative chronological marker (Birot and Joly 1952). It has also been proposed that the varnish can be used as evidence of certain environmental conditions. The rate at which it formed, which, depending on the authors, is estimated to be between a few decades to the (more probable) few hundreds or even thousands of years (Dorn and Deniro 1985), would have required some climatic stability. However, there are, as yet, too many uncertainties around the details of this stability and local conditions that may have prevailed in time and space. Thus, the varnish is not really a very reliable indicator of environmental conditions, except when it contains significant foreign bodies, such as identifiable living organisms, pollen or sand, carried by the wind. On the other hand, the varnish is important in the field of geomorphology, given its resistance and, especially, because of its silica content, as an effective protector of the underlying rock against aggressive erosion forces (especially the wind).

1.2.8. *Conclusion on weathering in the arid zone*

In the arid world, weathering is constrained by the uniqueness of the climate, as well as its irregularity. Climate determines the availability of water and variations in temperature, across time and space. The climate also determines the conditions under which such a process can occur and how frequently it occurs. In general,weathering decreases with an increase in dryness and at lower temperatures. However, these conclusions are also heavily blurred by the diversity of combinations between different microclimates arising, as a result of the relief, and by the resistance of the rocks themselves to the processes in question (Bourrié and Freytet 2008). Further, we must also add the nuances brought in by discontinuous green cover and the scattered presence of living microorganisms.

Chemical and biochemical weathering in the desert depends on the presence of water and the frequency of precipitation. It mainly occurs during the wet periods, which are spaced apart by dry intervals of varying lengths. The process breaks up and destroys the mineral edifices of the rocks and produces more or less complex *alterites*. These are either residues that are released and not carried away, or new combinations (neoformations) that stay in place or are carried away by infiltrating water. Given the scarcity of water and, in cold or temperate deserts, the very low temperatures, the process that prevails is usually a mechanical process (fragmentation), which occurs through the variations in temperature and moisture. This leads to the formation of *clastites*, which may be coarse or fine, depending on the nature and structure of the rocks involved. Massive rocks, with neither lamination nor a preferred orientation (intrusive rocks, limestone massifs), checkered by a network of joints and faults created by tectonic forces, form large debris, pebbles or blocks (> 10 cm), which are displaced by gravity and runoff. Upon disintegration, coarse-grained rocks that are formed of crystals (granites, metamorphic rocks) or debris of varying degrees of coarseness and of different origins (conglomerates, sandstones, pyroclastites), and which are compacted or cemented together, yield the grains that constitute them. This occurs in the form of gravel or granules (from 2 to 60 mm), sand (from 50 μm to 2 mm) or silt (from 2 to 50 μm), which are carried away by runoff or the wind. In stratified rocks (sediments, crystallophyllian rocks, volcanic lava), joints or stratification planes are often cut by fractures or faults. The resulting debris is generally kidney-shaped, needle-like or flattened platelets of medium size, gravels or pebbles (from 6 to 20 cm), which are spread through diffuse runoff. In loose rocks or formations, made up of isolated detritic elements, the progressive wearing away of the constituents eventually produces elements of very small size, fine sand or silt (<50 μm), which are dispersed by wind.

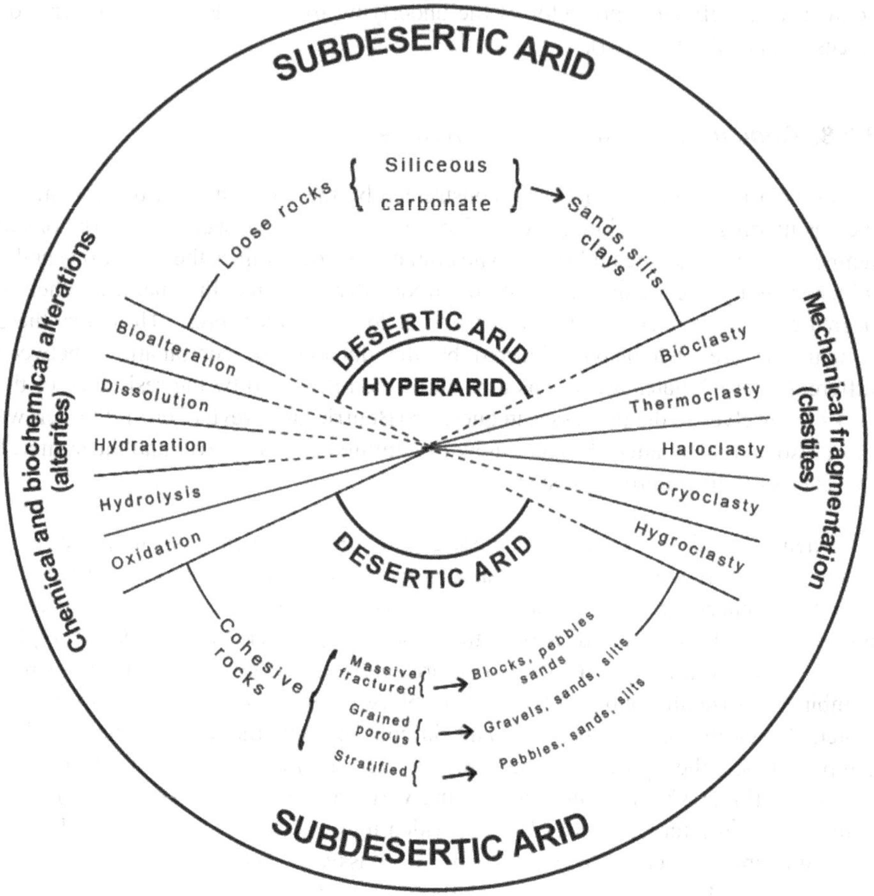

Figure 1.1. *Diagram summarizing weathering in arid conditions*

Weathering is especially present and effective in subarid and semi-arid regions in the hot zone (tropical and subtropical), where water is more abundant and better distributed. Nonetheless, the phenomenon of weathering is far from negligible in the regions with subdesert and desert aridity, although it is slower and more discontinuous (Figure 1.1).

On the contrary, weathering is highly limited or even non-existent in hyperarid regions, with very rare exceptions (temporary pools, rainy periods, humid recesses). The processes that require water (hygroclasty, cryoclasty, hydration, dissolution and hydrolysis) vanish almost entirely when reaching regions with desert aridity and hyperaridity. The only processes that occur in these regions (and even then, only in a highly attenuated manner with sporadic competition from living beings (bioclasty and bioalteration)) are those involving the circulation of air (thermoclasty and oxidation) or dust (wind polishing).

Chemical erosion and mechanical erosion contribute, at differing intensities, to rock weathering and fragmentation. There have been many attempts to analyze the exact contribution of each process (Bourrié and Freytet 2008); however, this is not easy to establish in an arid context given the highly diverse and scattered observations. Chemical weathering attacks the surface of bare or denuded rock, and then progresses toward the interface between the unweathered rock and the products of its alteration, which leads to a more or less rapid deepening of the alterite. In more humid regions, chemical weathering propagates more rapidly than mechanical erosion. Mechanical erosion occurs directly at the interface between atmosphere and the rock (or the weathering mantle that covers it), which is most often lacking in vegetation. It fragments the rock, and the denudation of the rocky outcrop that follows the removal of debris allows for new progress in the chemical weathering. While the process takes place more slowly on coherent rocks, it is however more widespread and, above all, more continuous than chemical weathering, which makes it more perceptible and certainly more spectacular.

Throughout the arid zone, the transformative and destructive action of weathering is slowed down and limited in time and space by the fleeting and discontinuous nature of the processes involved. Weathering has also varied greatly with the climatic changes in the Quaternary period. The processes taking place today over a decade or millennium also took place in the past, on a geological scale of tens or even hundreds of millennia. Dry and wet phases, and hot and cold phases alternated multiple times. Over the course of time, the processes of alteration and fragmentation acted upon different surfaces, with different results, but always in a relatively arid context. We must also recall that a large part of the superficial formations that we see today are legacy formations, and that present-day weathering is almost always being carried out on material that has already been shaped by the weather of the past.

1.3. Runoff

Runoff (R), or water flowing along the surface, occurs whenever evaporation (E) and infiltration (I) do not eliminate all the water (P) that has fallen on the soil:

$$R = P - (E + I). \qquad [1.1]$$

The *runoff coefficient* (C) expresses the ratio (percentage) between the quantity of water that flows away as runoff (R) and the quantity of water received as precipitation

(P) along a hillslope or a watershed:

$$C = R/P. \quad [1.2]$$

When precipitation falls onto impermeable or saturated soil, or when non-saturated soil cannot absorb the entire quantity of precipitation, the excess streams away and flows over the slopes under the influence of gravity. In the deserts, this run-off is always intermittent, fleeting in time and highly limited in space. It can take various forms, which are either aided or hampered by local climatic, topographic, edaphic and biological conditions.

1.3.1. *Diffuse runoff*

Diffuse runoff occurs when water from a downpour or from a spring flows in an unchanneled manner over a slope, without following a fixed channel that acts as a gutter. This is a disorganized runoff or stream, directly related to the arrival of rainfall (pluvial runoff) and/or when intensity of the downpour exceeds the soil's infiltration capacity (*overland flow*) (Horton 1933). Runoff is governed by a series of criteria that are particularly well represented in deserts: showers that are intense (greater than 0.5 mm/min) but far apart (at intervals of several months), relatively low slopes (roughly less than 15%), a discontinuous and thin covering of loose formations, and sparse and scattered vegetation. Diffuse runoff takes different forms based on the various possible combinations of these criteria and depending on the time and place. Given that it is essentially an episodic and highly circumscribed phenomenon, it is difficult to observe and measure. In fact, its observation often depends on chance.

An *elementary diffuse runoff* accompanies, or immediately follows, rainfall (*rain wash*). This runoff is made up of more or less transient pools, interconnected through rills, which snake around obstacles without growing deeper or more permanent. This *rill wash* is fed irregularly and generally disappears through infiltration or evaporation before reaching a larger, concentrated body of water. In the case of persistent rain, the shifting, braided rills can sometimes merge together to form a more or less continuous sheet that flows over the slope. One of the examples, observed by Joly on January 22, 1950 in the Moroccan region of the Sahara (Joly 1953a), a region with subdesert aridity, may give a better idea of this (*sheet wash*):

> After a dry summer, the winter of 1949–1950 was a particularly pluvial period on the glacis that stretched across the food of the Anti-Atlas, around the Aouin and Torkoz oases. These glacis were formed of large, flattened, jointed cones that came down the Jebel Bani. The slope was gradual and the vegetation was open.
>
> The weather had been quite bad from the evening of January 20, 1950 and a succession of stormy squalls, associated with a depression

> traveling over Morocco, had provoked multiple, violent showers. The sun was veiled; the invasion of the highly humid, maritime polar air had led to a considerable drop in temperature and, despite gusty winds, had reduced evaporation.
>
> On the morning of the 22nd, the soil was particularly soaked and it was difficult to retrieve the camels over the slippery mud of the *mader*[1] where they had committed themselves to. All through the morning the squally showers prevented us from packing up the camp. Towards noon, and lasting till half past noon, torrential rain produced the first sheets of water. Shortly after, around 12.45 p.m., a second downpour, even more violent than the first, lasted unbroken till 1.20 p.m., peaking twice or thrice with extreme force. It was this downpour that triggered a magnificent sheet wash over the glacis.
>
> Initially, the fine sandy formations between the scattered gravel absorbed the water till they were saturated. The shower was so strong that water could hardly penetrate the voids in the soil. Pools appeared, which were soon connected to one another through a complex network of thin streams that strayed away and skirted obstacles (a *rill wash*). Old channels, a few centimeters deep, were revived and quickly overflowed onto the flared banks. However, no new channels were created and there was no hierarchy established between the independent and shifting streams. The rain was incessant and the water levels rose in the pools, which joined up together and masked the initial, reticulated network. The flow then became a general flow in the form of a *sheet wash*, some 2–3 cm thick, which covered our sandals and flowed away downstream. This lasted ten to fifteen minutes. Finally the flood abated. The sheet of water broke up into isolated pools and the flow was once again divided into wide and shallow streams. Then, the rain stopped and everything ended abruptly. The waterlogged soil began to dry out again.

A similar phenomenon, but on a wholly different scale, was also seen on the alluvial plains and fans, through sheets of water resulting from a flood in the rivers (*sheet floods*). For example, in the alluvial plain of Mahjez, in the middle Daoura, at the end of the gorges through which the Wadi runs in the Kem-Kem region (Algerian-Moroccan region of the Sahara), in a region with desert aridity, Joly made the following observations during a rainstorm on February 14, 1949 (Joly 1953a):

> The rain quickly wet the clayey silt of the fan, which had been fissured by the drought. The cracks closed up and the entire surface became

1 *Mader*: intermittent agriculture.

smooth and impermeable. This wet layer, which was so thin that the footstep of a man or a camel could reach the dry soil below, allowed for some surface flow. The streams on the surface traced little furrows that converged into deeper channels, flowing along the uneven ground, but growing no deeper (*rill wash*). As the rain persisted, it combined into sheets (*sheet wash*), which rushed into flood channels, smoothing away the angular edges of the banks. At the same time, the flood flowing into the wadi from upstream, first flowed into the channels created by the last flood, and then overflowed into sheets (*sheet flood*) and spread out into a fan, absorbing all other flows into itself. This combined flow then swept across the alluvial fan. When the flood abated, on the other hand, the runoff flow concentrated into a smaller number of gutters, which were more or less deeply etched into the alluvial mass.

1.3.2. *Concentrated runoff*

Concentrated runoff occurs when rainfall or floodwaters flow in a channel. The quantity of runoff is measured by the *modulus* or *flow rate*, in cubic meters per second, evaluated at a specific point in the course of the flow. To allow a better comparison between rivers whose watersheds have very different areas, wherever possible (since the area of an arid catchment basin is not always easy to measure), we use a *relative modulus* or *specific flow rate*, which is the ratio of the flow to the area of the watershed, expressed in cubic meters per second, per square kilometer. This ratio is usually quite small for large basins in arid regions. Compared to a diffuse flow, a concentrated flow corresponds to precipitation of the same order, but that is more spread out in time and space, and generally falls over steeper slopes. Although not easy, it is possible to study concentrated flow using the same measurements and models usually used for other water bodies (Cosandey and Robinson 2000).

Apart from "humanized" streams and large rivers, fluvial hydrology in the desert is a field that is still poorly understood. Indeed, qualitative observations and, sometimes, highly precise measurements have been carried out in most places, either in the field or in an experimental setting (Braquaval 1957; Schick 1970; Savat and De Ploey 1982), but they remain isolated observations, forming a series of examples rather than a consistent set of results. It is still possible, however, to identify essential features from these observations (Dubief 1953; Joly 1968; Schick and Lekach 1987; Rodier 1982).

Along the subarid and semi-arid borders, rains may fall annually or seasonally but they are always weak. Runoff streams are generally intermittent but may sometimes be permanent. In zones that are truly arid, irregularity is the norm; runoff streams are always intermittent and are sometimes transient. In hyperarid regions, runoff is always transient and random. The only real water-holding sites of any significance are mountains or relief features that are large enough, which can cause rains resulting from convection and storms, where the slopes are steep and the well-marked traces of past flows facilitate the concentration of water. But nowhere in the arid regions are there local, perennial streams. The arid water body *par excellence* is one that is essentially episodic in nature: the *wadi*. In a wadi, flow and floods are often the same. In some regions, the same word is used for both phenomena. Flowing water is guided by banks and its volume is augmented by sheets and rills that flow down slopes. The delay between the rain and the flood varies depending on the distance, the flow rate, the slope and the permeability of the bed of the water course. There are two main kinds of floods, as follows:

– *Local floods* may occur at any point along the course of a water body. They are due to concentrated rains over a limited space (some tens of km^2). As in the case of diffuse flow, a flood occurs after a heavy shower and on soil that is already wet. The permeable alluvial material that lines the bottom of a talweg takes just a little longer to reach saturation. The distance the flow travels increases with the intensity of the rainfall and then the waters go down and disappear into the alluvial deposit itself, or evaporate when the sun reappears. Here is the description of a flood of this kind, observed in December 1949 in a small wadi in the Daoura hamada (the Algerian-Moroccan region of the Sahara), in a region with desert aridity (Joly 1953a):

> The rain first appeared on the morning of December 1, in the form of a persistent drizzle, wetting the soil along the slopes and the beds of the wadi without, however, producing any real flow. The rain stopped in the afternoon, but heavy, loud clouds continued to travel across the sky under a westerly wind. The rain reappeared more violently in the evening, between 7 p.m. and midnight, coming down in heavy, brutal showers. This time all the gullies and ravines had water flowing through them. Pools appeared in the bed of the wadi itself around 9 p.m., formed both by the rain as well as by the emergence of underflows. As these pools grew, they merged with one another to form a long pond (*guelta*), which was animated by a weak current. Towards midnight, as the rill-washes descended the slopes, a large wave, about some ten meters or so wide and between 1 and 1.5 m high, entered the talweg, moving at a speed of 1 to 2 m/s and fed by the overflowing ravines and drainage channels in the glacis. The flood abated almost immediately after this, although the rain continued in the form of drizzle. But the waters receding became evident only around 6 a.m. Between 7.30 a.m. and 8.15 a.m., the water level decreased rapidly and, in stony banks, went back to flowing in canals. When we left the site, around 10.00 a.m., we were able to walk without our feet getting wet. The talweg had hardly changed: at a few places the banks had collapsed, it was

a little sandier in some spots and had fewer stones in others. This was the only overall effect the flood had.

– *Remote floods* arise due to rains that have fallen on relief features located some distance away from the site of the floods, with the precipitation having occurred some time before the arrival of the influx. These floods always travel over a significant portion of the course of the water body, sometimes traveling its entire length. They may suddenly arrive at a site that has had no rain at all. To be drowned in a desert is not just a far-fetched figure of speech. Observations from October 1950 in the southern Algerian-Moroccan region of the Sahara amply illustrate this scenario (Joly 1953b; Vanney 1960):

> September and October 1950 were particularly rainy across the southern slopes of the High Atlas mountains and in the Anti-Atlas. On October 12, around 6.30 p.m., there was a large flood in the Saoura that passed through Kerzaz, some four hundred kilometers further south, which was mainly formed by rainfall (around 100 mm) that fell from the 6th to the 8th in the High Atlas basin of Guir. On the morning of the 13th, the flood had swollen further. The wadi, which was very broad, flowed at the foot of the western Grand Erg, carrying yellow waters, heavily laden, with loamy flakes floating on the surface. The waters ate away its banks, carrying away great chunks of the dunes and of silt in one fell swoop. The same day, around half-past noon, roughly 18 hours after the flood had passed through Kerzaz, the floodwaters reached Foum el Kheneg, 80 km downstream. A breach opened in a sheet of silurian sandstone allowing the water to travel towards the neighboring depression of the Melha sebkha. The head of the flood thus progressed slowly enough (5 to 6 km/h) for us to be able to follow it on foot. The water flowed in channels created by the previous flood, destroyed dunes that had formed in the talweg and washed over the culvert in the path in a cascade. From this point onwards, the flow divided into two branches, as has already been described by E. F. Gautier (1946): one route, usually followed by "the usual floods", merged with the Melah sebkha to the north, while the other route, called the Messaoud wadi, was the outlet for exceptional floods, and led away south to the Touat oasis. On the evening of October 14, at Béni Abbès, 200 km further north, the flood in the wadi abated but was still too wide and tumultuous for the wadi to be crossed on foot. We could see it washing over the culvert on the road, accumulating deposits upstream and digging deep cauldrons downstream, in which the water formed violent whirlpools that weakened the foundations of the culvert. Alluvial deposits formed protrusions along the submerged floodplains and the height the water reached was marked, in many places, by the creation of a small line of erosion in the silt of the cliffs. The floodwaters continued to go down all through the day on the 15th. Around 5 p.m. that evening, a large breach opened up in the culvert. The water level

could then plainly be seen to fall and the flow was concentrated into four or five channels, upon which the emptying streams converged, indenting the floodplain. The flow was, thus, sustained for several days, until a fresh flood arrived.

In the winter of 1950–1951 this region saw a number of remarkable flow and erosion phenomena, some of which reached catastrophic proportions locally. The Dra wadi, which is often dry by the end of summer, or only traversed by some storm-fed waters, had a continuous flow of water from 9th August onwards. On October 4, muddy waters from a flood that was dying away washed over the Zagora culvert bridge, but a new flood had already formed in the north, moving at a speed of 10 km/h. In the Tafilalt depression, the Ziz and Rheris wadis were also flooded, fed by the abundant rain falling in the High Atlas and the Djebel Sahrho. On October 5, all the wadis in the basin had water flowing through them. When the flood died away, the fall in water level was immediately noticeable and relatively even (10 to 15 cm/h), before becoming slower. The water of the Rheris, after having passed through several palm groves, was bistre (grayish-brown) and heavy with silt. The waters of the Ziz, on the other hand, which had just flowed over the Cretaceous limestones of the Meski hamada, were clear. After the Ziz and Rheris had overflowed onto their floodplains and filled up the neighboring *maders*, they merged with the Daoura wadi which in turn, began to flow from October 7. On October 16, the local floods were still reviving the small wadis in the Guir hamada, and the surface of the plateau was punctuated by temporary pools. During this time, fresh, intense rainfall over the Atlas triggered heavy new floods (1,000 to 1,100 m^3/s) in the still-waterlogged talwegs of the Rheris and Ziz, which would cause significant damages in the Tafilalt basin.

These observations are only examples and different events may occur during other floods. These depend chiefly on the force of the flood, which may arrive with great ferocity (the oft-described "wall of water") or completely exhausted of all its power. They also depend on the distance that has already been covered by the flood, on the volume and the nature of the suspended load the water carries (solid flow rate), or again, on the state of the talweg when the flood reaches it: bare or covered with coarse or fine alluvial deposits, whether or not there is a border of vegetation, dry or already wet from an earlier flood.

In his novel, *Madman's Bend*, A. Upfield (1963) describes a flood in the Darling, the "Australian drainage basin" in southeast Australia:

There was nothing spectacular about it [the arrival of the flood], merely a carpet of leaves and debris carried by a low water surge. Just below

> *Madden's homestead the carpet appeared rapidly to thicken, rumple, break into separate pieces; and these moved faster until they rejoined and the carpet was re-laid [...] behind it further debris was added to the mass, which again piled upon itself until the pressure broke it into smaller masses of leaves and bark streamers and small branches. Thus the river methodically built its dams and destroyed them, the water not deeper than fifteen inches [...] The water was the colour of putty [...] the free floating branches and tree trunks which moved relentlessly past [...] were gently pressed against the barrier, thickening it and apparently strengthening it. Before [...] the water had been only knee-high; now behind it the water rose to the height of a man [...] [then] rose swiftly to fifteen, eighteen, twenty feet. [...] The stream was rounding the bed ponderously, bearing further trunks and branches.* At the peak of the flood: *There was nothing of turbulence, and yet nothing of gentleness in the flow. It was not unlike tarnished gold on its way to distant moulds, and the odd pieces of flotsam seemed to be partially buried in half-solid matter [...] Three nights and two days passed before the river began to fall.* And from then on: *Its speed was about the same, but [...] it would gradually become slower as the lakes and billabongs and creeks lower down filled, and the general course would be broadened at places for miles.*

See also Gale and Bainbridge (1994).

The most remarkable floods often result from the combination of a remote flood and one or more local floods. This is when we get the best observation of the river in action.

When it arrives, the floodwaters are heavily loaded, to the upper limit of its transporting capacity, as the flood encountered an excess of mobile material in the talweg. The upper layers of the water, which are less dense and faster-moving, surge up on the lower layers, which are heavier and slower. As a result, the waters retain enough of their eroding ability, which can only act laterally. Thus, the banks and bordering levees are first attacked; loose formations along the floodplain collapse entirely. Whirlpools are formed at the height of the flood. Some of these, with a horizontal axis, stir up the alluvial deposits at the bottom but do not dig through them, or to a very small extent and in an uneven manner. Other whirlpools, with a vertical axis, caused by the difference in speed of the different currents of water and the roughness of the banks, contribute to the erosion of the banks and the widening of the riverbed. When the flood overflows onto the flood plains or expands over an area of deposits, it loses power and transforms into a *sheet flood*. It is often at this point, as the flood abates, that vertical erosion is most noticeable. The partial deposit of the load carried by the water, and the concentration of the flow, reactivate some of the water's eroding capacity, resulting in a limited number of channels in the alluvial deposits in the regular stream bed; lower water levels trigger an eroding wave that moves backwards through the emptying ravine in the floodplain (Joly 1953a).

In agricultural zones, a flood is both welcomed and feared: it brings in new alluvial deposits and recharges the groundwater tables. However, it also carries away gardens and trees and can also lead to irreversible transformations in the morphological landscape, which can disrupt life for local populations. Another example from the Tafilalt basin (southern Morocco) illustrates this (Joly and Margat 1955; Joly 1962).

> The Tafilalt plain is a meridional topographic depression, lying on the saddle of the major tectonic axis (the eastern end of the Anti-Atlas). The plain itself is made up of Quaternary deposits, the majority of which are sandy-silty, that form a layer that is about 15 to 30 m thick over a folded Paleozoic substrate. The plain is formed by the convergence of two rivers originating in the High Atlas, the Zis and the Rheris. However, this convergence is not accompanied by a confluence of the two wadis, although they are only separated by 6.5 km around Erfoud. In their present courses, the Ziz and Rheris flow separately in two distinct riverbeds, with a relative difference in levels, the Rheris flowing 5 to 12 m below the level of the Ziz. To the south of Tafilalt, the two wadis diverge to go around the small Taouz massif, before finally reaching a point of confluence to form the Daoura. However, this was not always the case and older points of confluence existed, as can be seen through ancient traces (beds of round stones) that are abandoned today, covered with sand and silt, cultivated into gardens and used for irrigation canals (*seguias*). The silty plain may be viewed as an alluvial plain common to the Ziz and the Rheris, with the coarse alluvial deposits coming mainly from the Ziz, which is closer to the mountain, with the Rheris being decanted downstream by a string of South Atlas plains and carrying largely sand and silt. Until a recent period (the end of the Pleistocene), the Ziz extending beyond the Tafilalt made up the lower valley of the Rheris. However, over the course of the infilling, several discharges from the Ziz-Rheris, facilitated by the convexity of the surface, led to a small local wadi (the Amerbouh wadi), which was situated to the east of this plain and had a Southeast orientation. The formation of a trough carved out around the start of the Holocene epoch (occurring no doubt during the shift from climatic optimum to aridity), led to the lasting separation of the two wadis and allowed the development of gardens and the large Erfoud and Rissani palm groves in the space between them. The present surface of the silty plain is, in fact, an anthropized surface; several meters of the infilling material were eroded and the primitive

surface subsists as isolated knolls. Indeed, the course of the Ziz through the palm groves (the Ziz Filali) is an artificial course, which is used as the main seguia and is also the route taken by floodwaters. Some of these floods, like those that occurred in 1950, which are spread out across the plain or channeled into old canals, benefit from the difference between the levels of the Ziz and Rheris, creating a confluence that benefits the Rheris. This *overflow capture*, followed by intense regressive erosion in the inflow channels, tends to isolate the entire downstream part of the plain and destroy the palm groves set up here, unless there is human intervention to counter these effects artificially. These kinds of crises demonstrate how an anthropized zone, such as Tafilalt, is constantly threatened by natural geodynamic evolution. By re-establishing a principal flow consisting of the Ziz filali each time, to save their crops, humans make an already risky equilibrium all the more precarious.

1.4. Geodynamic and geomorphological effects of runoff in arid zones

The geodynamic and geomorphological role of moving water is primordial in the desert (Scoging 1989). There is no region, except perhaps the ergs (and a few other exceptions), which has not witnessed the phenomenon of runoff at some point or another. The essential action of the runoff is not really very different from that in more humid parts of the globe. The processes brought into play are the same, however they are distributed differently through space and time and, above all, their effects are slower as they are intermittent (Joly 1953a; Joly 1962).

The major activity of running water is to displace detritic material produced by the fragmentation and alteration of rocks through processes such as tectonic movement or weathering. Hydraulicians are able to use equations to establish relationships between the different terms involved in this work, but the numerical data are very often lacking. To simplify things, let us say that the liquid current must possess a certain quantity of energy (*power*, w), which depends on its volume and its speed, to transport a certain quantity of debris (*load*, or *solid flow rate*) determined by its mass (*transport capacity*) and the quality (*competence*) of its constituents. If the power, w, is enough to displace the load l, *erosion* takes place through the ablation of material and through the friction against the walls of the stream bed. On the other hand, if power diminishes in a loaded current, then there is *deposition* (or sedimentation) of the alluvial material in transit. For a given current, this fragile equilibrium depends on climatic conditions (the quantity of rainfall, weathering), on lithology (nature and resistance of the rocky material, the supply of the load), the topography (the slope values and flow speed), as well as on earlier contributions (previous floods) or external contributions (inflows) and losses (infiltration, evaporation, pumping).

In hyperarid regions and regions with desert aridity, the major mass of debris covering slopes (blocks, gravel, sand) is produced through mechanical fragmentation. Finer material (<2 mm) is swept everywhere by the wind. On steeper slopes (>20°), especially in mountains, at the foot of craters or the edges of highly fractured plateaus, the liberated fragments have a tendency to migrate downwards, even in the absence of water, under the influence of gravity alone. The movement of this debris along the slope becomes easier as steepness increases, when the fragments are larger and no obstacles (plants, for example) break their fall. These fragments then form *screes*, which may be regularly carried away downstream, failing which they form permanent structures and eventually fossilize the hillslope. Indeed, truly functional screes are quite rare in the desert, or are highly localized, as the poor erosion of the steep slopes does not supply fresh material, or does so very slowly, and on low downstream slopes, they can only be displaced by run-off. Along the the small slopes (<15°), formed by glacis or flattened cones covered with gravel and sandy deposits formed by the dismantling of fossil crusts or brought in by wadis, gravity alone is not sufficient to move the thick debris along the surface. This is moved only through the diffuse run-off that follows intense showers.

The flow forms directly on the bare rock, or forms more slowly if there is a more or less continuous covering of permeable loose formations that can absorb the water. The arrival of the rain onto saturated soil leads to the formation of pools of water in hollows, and streams that weave their way between various obstacles (tufts of vegetation, rocky blocks, creases in the earth or sandy ridges).

Former canals, which are a few centimeters deep, are revived and start functioning again, soon overflowing their flared banks and forming a *sheet wash.* The permeable layer and fragments of different dimensions that cover the soil make up a potentially heavy load of debris, depending on the flow's strength. Initially, the flow is neither clear nor muddy and only carries particles the size of sand. The flow shows no ability to carry pebbles, but they may sometimes be displaced when the fine debris that supports them is carried away. At this stage, no hollowing out is seen. On the contrary, the sand accumulates in long walls or in microdeltas that attach to small obstacles, mineral or plant, but no new incision appears and there is no hierarchy among the rills. The fact that during the flow we often see deposits but never any hollowing away, suggests that the flow is almost always completely loaded to capacity, which would explain why the flow cannot become concentrated. The flow becomes concentrated toward the final phases of this phenomenon, when the sheet breaks up to form a few streams. However, the flood ends before any deep hollowing of the ground can take place. In these conditions, *erosion* or *denudation* (H. Baulig) takes place over the area covered by the sheet: *areolar erosion* (P. Birot). After the flood dies away, its course is paved by large debris and multiple channels, more or less anastomosed and barely marked, filled with fine debris, similar to the channels left behind by the previous *sheet wash*.

The morphological expression of channeled running water is the canal. It follows a linear path (*talweg*) that is the site of the lowest points in its course. There are canals of all sizes, ranging from the basic *gully*, which is relatively shallow with sparse branching, to the water courses that run over several kilometers and are organized and arranged into a hierarchic network of varying complexity (*hydrographic networks*) and varying levels of activity, fed by atmospheric precipitation, diffuse run-off from the slopes and/or by springs and upsurges of underground water (*aquifers*).

1.5. Groundwater

Some part (a varying quantity) of the water that comes from the atmosphere is absorbed by soil and stored at depth. In deserts, this is an important process, both hydrologically as well as biologically, as it removes a significant quantity of life-sustaining water from surface flow, from plants, animals and humans, and is not always recoverable. This process of water being stored as groundwater is subject to various factors, which are not all equally active in a desert environment.

1.5.1. *The penetration and circulation of water at a depth*

Water that reaches the soil and flows on the surface penetrates to a depth by substituting for air in the voids within soil and rocks, doing this at varying speeds and more or less completely. This is the problem of *infiltrability* (Cosandey and Robinson 2000). In an arid zone, a slow and prolonged shower, or a heavy shower falling on a loose formation with many voids open to the exterior, may infiltrate directly, without any run-off. On the other hand, the rain may beat down with such intensity that the soil exceeds its absorption capacity and the water runs off without infiltrating. Further, unlike humid regions, the desert has large areas with bare or denuded rock and hardened surfaces (crusts or -cretes[2]) with low permeability; water therefore glides off these surfaces without penetrating them. At the same time, the sparse and scattered vegetation reduces the role of infiltration along the roots. In general, water largely tends to run-off, especially if the slope is steep, there is heavily concentrated rainfall, there are rocky outcrops, and scattered vegetation. On the contrary, a layer of loose mantle, soil *sensu stricto* or surface formation and/or denser vegetation, especially herbaceous green cover, acts like a buffer layer, and their presence helps infiltration take place. Water travels easily across a loose or pebbly

2 -cretes is used in words such as calcretes, ferricretes, silcretes to designate crusts made of calcite, ferric oxides or silica, respectively.

cover, as in the regs formed by fragmentation, or the alluvial deposits in talwegs. A thick, sandy-silty cover forces the water to travel slowly from the surface to the rocky bed – this can be as much as 90–95% of the water that fell, and run-off is deficient here. However, across the silty or clayey alluvial plains, less than 25% of rainwater infiltrates the soil and the rest contributes to smoothening the surface and making it impermeable. In fact, the greater part of infiltrated water comes from diffuse run-off, floods in the rivers and losses by temporary pools and lakes, rather than coming directly from rainfall (Dubief 1963). This infiltration is also limited in quantity by a factor called the *infiltration capacity*, which depends on both the quantity of water fallen per unit of time, and the external openings of the voids in rocks, i.e. pluviometric conditions and the state of the surface determine the penetration of water. However, once water has penetrated, its transfer and storage is governed by the structure of rocky material, inherited from lithology and tectonic history.

The key feature here is *permeability*, the aptitude of a rock or geological formation to absorb and allow fluids, especially liquid water, to circulate. Permeability depends, first of all, on the rocks themselves. In coherent rocks, either stratified or fissured by tectonics, water penetrates and circulates easily since the voids (joints and fractures) are wide and interconnected. When the cracks are very narrow, they are quickly clogged by impurities and the rock becomes impermeable. On the other hand, if the rock is easily weatherable, as is the case with crystalline rocks, or soluble rocks such as limestone or gypsum, the cracks continue widening until they form proper underground networks. In this case, we speak of *large-scale permeability*, where water passes through easily but is not stored. In loose rocks, like sand, or porous rocks like sandstones, penetration and circulation of water depends primarily on the void dimensions and their interconnections. Screes and coarse alluvial formations allow water to freely penetrate and circulate. Sands, silt and porous rocks absorb water more slowly but retain it for longer. This is *small-scale permeability* as small voids are involved. Plastic rocks and clays in clayey deposits, or in sebkhas, are fissured and permeable when dry. However, with water entering the shrinkage cracks, they soon transform into an impermeable colloidal gel. These are very general conditions, not specific to arid zones. However, permeability also depends on the quantity of water held from previous infiltrations, as saturated soil is impermeable. In deserts, water is quite rare and unless there are a series of showers, rainfall occurs at intervals of several months, if not years. Thus, the "last infiltration period" is always in the distant past and the soil is most often dry. Consequently, the infiltration capacity is almost always maximal. A particular case in deserts, common in cold deserts, is that of frozen soil, occurring either seasonally (*gelisol*) or permanently (*pergelisol* or *permafrost*). The frost, which blocks up the voids and fissures in the rocks, makes the soil impermeable for the entire duration that it is frozen. On the contrary, in spring, the alternating frost/thaw cycles lead to fragmentation and fissuration in rocks (frost weathering), thus increasing their permeability, which in turn leads to an increase in the absorption of rainwater in the summer. On the other hand, the alternating episodes of snowfall and the slow melting of the snow, toward the end of the summer and in autumn, impregnate the soil to a significant depth and to a much greater degree than the rapid summer phenomena.

The infiltrated water moves from top to bottom, and obliquely through the voids in the rock under the influence of gravity. However, this movement is counteracted by evaporation, an ever-present feature in the desert, and by being sucked up through plant roots, if they are present. Evaporation starts as soon as rain falls on the soil. Soil moisture decreases with depth but can still be noticed over an entire layer of the soil, not exceeding 3–4 m in thickness, however. This is an aeration and diffusion layer where infiltrated water is partially retained through adsorption onto certain hygroscopic minerals like anhydrite, clay minerals or iron and aluminum hydroxides; it is retained through capillary action in smaller voids. The upper part of this layer dries out very quickly, as soon as the input of rainwater or run-off water ends. Water remains present for a longer period in the lower part, where it is transferred until it encounters an impermeable layer (aquitard).

1.5.2. *Groundwater reserves*

Upon arriving at an impermeable layer (aquitard), the infiltrated water accumulates and forms a *water table*, stored in an aquifer (Castany 1967; Castany 1982). In a desert, these groundwater reserves are very precious as there is more water available here than on the surface. Above all, there is greater certainty with this water, which is not as inconstant as surface waters and is longer lasting. Unfortunately, groundwater cannot always be reached by those who may truly need it. There is considerable difference in arid zones between different groundwater reserves, based on their hydrogeological setting, their chemical composition and their recharge (Loup 1974).

Phreatic groundwaters are reserves that are not located very deep in the soil and that remain in contact with the atmosphere through the voids of the formation in which they are held. Their upper surface, the *hydrostatic* or *piezometric* surface, is a capillary fringe with variations in the level of the water table. In humid regions, phreatic groundwaters receive water mostly from rains and these reserves maintain the permanent flow in rivers through springs. In the desert, conversely, rivers mainly feed groundwater tables and these tables only sustain baseflows when water levels dip in talwegs. The most significant of these are the *underflows*, as they are the most accessible. These are water currents that circulate in the alluvial deposits in riverbeds, sheltered from excessively strong evaporation. The underflows are fed by floods and, just like the floods, they are intermittent. However, they are propagated more slowly, persist for longer after the baseflow, and can sometimes last from one flood to the next. Most importantly, they go further downstream and sometimes, unlike the floods, the underflows may even travel the entire length of a talweg. This occurs, for example, in large wadis such as the Saoura (arising in the Atlas) whose underflow can sometimes extend up to Touat. This is why hydrographic networks in the desert are so important for the survival of living beings.

The water that flows in rivers also infiltrates the sides through the banks and feeds the lateral groundwater tables, which also receive infiltration water from rainfall and diffuse run-off from slopes. The extension of these water tables is governed by the magnitude of the floods and is limited by the dominant relief features. When there is a powerful river (usually allogenous, i.e. coming in from another region) with large floods and flat topography, the water table may be considerably large or widespread. Loup (1974) points to the example of the Murray (to which we can add that of the Diamantina) in Australia, whose lateral water tables extend over tens of kilometers on either side of the river, as well as the example of the Nile in Egypt, which feeds the water table and the lakes of the Wadi el Natrun, about 30 km to the northwest of Cairo. Other lateral water tables, of various extensions, accompany rivers that descend in mountains, like those around the Hoggar, the Aïr and the Tibesti in the Sahara. The drawback with these phreatic groundwater reserves, however, is their high salt content, resulting from the extended period they are held in these *reservoir rocks* and the concentration of these salts by evaporation.

Deep groundwaters (Margat and Saad 1984) are more remote from contact with the atmosphere. Many of these deep groundwater tables can still receive rainwater and water from surface flows through infiltration. However, they are recharged very slowly, if at all. On the other hand, they lose almost no water through evapotranspiration and, consequently, form a stable reserve constituting a true "mine of water under the desert" (Margat 1983). It is the same with massifs, with schists, or other crystalline rocks that are porous or fractured by tectonics. Paradoxically, dunes, which are the most popular symbol used to represent waterless deserts, are among the best reservoirs of water in the desert! Water travels easily through the sand and accumulates on the rocky floor that lies underneath it. All the large ergs contain deep water tables (Cornet 1952). However, these are not easily recharged given that, in the absence of wadis and floods, the only source of water is from rainfall. Sand is a good filter and traps small solid impurities, and when the sand is made up of siliceous grains that have very low susceptibility to alteration, the non-salty water in these reserves is often of excellent quality. In large sedimentary basins, where a permeable aquifer is intercalated between two impermeable layers (aquitards) the water table held here is said to be a *captive groundwater table*, separated from the atmosphere by the dry layers or other aquifers superposed onto it. If the captive water table has no outlet and is only fed and drained by seepage through fissures in the layers that surround it, then the water is not recharged, or, if it is, is recharged very slowly. In these cases, we use the term *fossil groundwater*. The oldest of these reserves could, theoretically, date back tens of millions of years. In fact, however, the water is usually much more recent and, based on isotopic analyses, most of the fossil groundwater available today seems to have been recharged during the wet periods in the Quaternary, especially during the last large interglacial and the Holocene climatic optimum. If an aquifer layer is reshaped by tectonic movement into a monocline fold or a syncline basin and is laterally transected by the topographic surface, it remains in contact all along the outcrop. In the water supply zone, the hydrostatic level of the table remains free and fluctuates. However, in the captive regions, the water is constantly under pressure. At all points on the soil surface located at an altitude lower than the hydrostatic level, the groundwater may spill out through fissures or cracks in the upper layers, or rise up through wells. These are called *artesian aquifers*.

A special case is that of *karst reserves*, found in limestone or gypsum environments. These are quite rare in deserts, not because of a lack of appropriate rock but rather because of the scarcity of water. When these do exist, they make use of groundwater reserves inherited from older, more humid periods (largely Plio-Pleistocene), which are traversed by random and fierce floods. The largest of these are usually those where the water supply zone extends beyond the desert (mountains or regions with high pluviometry), and/or those that benefit from infiltration from water bodies or perennial lakes. Some of these are artesian aquifers and feed springs that emerge along fractures or fault-lines, such as the Ras el Aïn springs, for instance, which emerge along the semi-arid limestone plateaus in northern Syria and which are considered the largest karst springs in the world (38.6 m^3/s), or the Al Hofuf springs along the eastern coast of Saudi Arabia (14 m^3/s) (Nicod 2000). Other karst networks exist in the Sahara – in the Toudeni basin, for instance, the large Holocene lakes were fed by karst springs (Riser 2000) – Australia (Nullarbor, on the southern coast), New Mexico (the Pecos River, south of Santa Fe), South Africa or again in Madagascar (Salomon 1987, 2000).

In an arid environment, however, these reserves are not very reliable as they are more similar to streams (episodic and transient) than to groundwaters. However, when they do work, they play a role in regulating floods by slowly restoring water collected during rains and storms, and feeding longer baseflows.

1.5.3. *Large hydrogeological basins*

Large underground hydrogeological structures are not exclusive to desert regions: these formations depend on structural causes rather than climatic causes. However, in the desert these structures are marked by aridity and take on special significance as they constitute the most abundant reserves of water in these regions (unfortunately they are also the hardest to access). All of them are located in regions where the ancient basement (Pre-Cambrian or Paleozoic), folded and/or fractured, has been depressed into basins with greater or lesser subsidence and covered in unconformity by sediments (mostly of continental origin), which act as receptacles for water and also often for other fluids such as gas or oil (Sahara, the Persian Gulf, the Caspian sea). These kinds of systems with deep groundwater exist in most large deserts (OSS 1995).

The most famous example of this is the hydrogeological basin in the septentrional Sahara region (Algeria, Tunisia, western Libya). The region occupies a vast Meso-Cenozoic syneclise[3] that dips northwards from Tademaït and Tripolitania up to the south of the Atlas, passing through the great Tunisian salt lakes. The existence of this basin on the surface has been known for a long time and has been exploited as well. However, its complex structure was discovered due to oil prospecting carried out there since 1946. The whole basin is made up of several superposed artesian aquifers, forming a total reserve measuring 60,000 billion cubic meters (Margat 1983; Mutin 2000), divided into two groups. The oldest and deepest (800 – 1,500 m) is the "Albian aquifer", which stretches across an area of around 600,000 km^2 in the sand and sandstone formations of the Lower Cretaceous Continental Intercalaire. This is a fossil groundwater aquifer containing hot water (55 – 60°C) that is also salty (up to 6 g/L) with very slow peripheral recharge (that is also not well understood) of the order of around 270 million m^3/year. The most recent group consists of four or five tiered aquifers, over around 350,000 km^2, contained in sands, sandstones and limestones in the upper Cretaceous and Mio-Pliocene Continental Terminal in the hamadas. These are aquifers with cooler water (30 – 35°C) that is less salty (1.5 – 6 g/L), which receive around 580 million m^3/year and are more accessible as they feed several emerging springs.

A similar basin, but even larger (around 2,500,000 km^2) and more complex, exists in the easter Sahara region, from Libya to Egypt and to the north of Sudan (Mutin 2000; Sanlaville 2000). The sedimentary aquifer series extends from the Cambrian to the Upper Cretaceous period and mainly consists of continental sands and sandstones, called "the Nubian sandstones", interstratified with less permeable marine deposits. These formations occupy a stratigraphic position comparable to that of the Continental Intercalaire in the western Sahara region. The total capacity of this reserve is around 50,000 – 60,000 billion m^3 and is recharged overall by around 1 billion m^3/year (Margat 1983). This entire structure is divided into several secondary basins separated by structural domes that also communicate with each other. To the north, along the Mediterranean and in the Kattara basin, the water table runs very deep (2,000 m) and is overlaid by sand aquifers from the lower Miocene, which interface with seawater west of the Nile Delta. Salinity can be as high as 5 g/L. To the southwest, in Libya, the Koufra basin is less deep (800 m) and has lower salinity (0.15 – 0.25 g/L). To the center, the Dakhla basin in the Egyptian desert, extending as far as the Nile valley, is the largest of these secondary basins. The Nubian sandstone, which is very thick (1,500 m) and divided by impermeable

3 Syneclise: a large fold in a syncline that can be up to a hundred kilometers long.

levels and secondary tectonic accidents, contains several artesian water tables, which are the least saline in the region (0.1–0.8 g/L). Further, this aquifer is very close to the surface in the large depressions in the Egyptian desert: Dakhla, Farafra, Kharga. To the southeast, the Egyptian-Sudanese Upper Nile basin includes (on a thickness of some 500 m) schists and sandstones dating back to the Upper Cretaceous period, and limestones dating to the lower Cenozoic. This groundwater is not as deep and is better fed by the floods in the Nile, discharge from lake Nasser, tropical rains and infiltration along several brittle tectonic deformations. Other deep groundwater tables, which may or may not be artesian, occupy practically all the large depressions in the Sahara (Fezzan and Tanezrouft, for instance), in the Continental Intercalaire, as well as the South Atlas hamadas in the Cenomano-Turonian limestone (Mzab) and the Continental Terminal (the Hamada du Dra, the Guir and Bechar hamadas, and the plateau of the dayas).

In the Arabian peninsula (Sanlaville 2000), which is a structural and climatic continuation of the eastern Sahara, deep aquifers exist in the sedimentary series (laying in the south and the west on the mountains of Yemen, Asir and Hejaz) dip down to the east and the north toward the Persian gulf, Jordan, and Iraq. The deepest of these aquifers are the captive fossil groundwaters held in the Paleozoic and Triassic sandstones, in aquifers around 2,000 m thick that provide hot water (38–57°C) with low salinity (0.1–1 g/L). These aquifers still receive rainwater and floodwaters through the outcrops to the west and south that are located in higher regions with higher precipitation. The upper groundwater tables are not as well supplied with water, as their outcrops are located in the lowest and most desertic part of the peninsula. Held in Lower Cretaceous and Middle Cretaceous sands, and in Eocene limestone and sandstones, these groundwaters are cooler (35–40°C) and generally saltier. The most recent of these outflow into sebkhas and are discharged through continental or undersea springs.

Further east is the large hydrogeological network in the Turanian or Aralo-Caspian depression (Turkestan). Here again it was oil prospecting that led to a greater study of the characteristics of an aquifer comparable to the large Saharan aquifers. The principal groundwater reserve in the Aral basin (Létolle and Mainguet 1993), lying on the Uzben mountains to the southeast, dips and stretches toward the northeast, as far as the Caspian sea. This groundwater is held within the Upper Cretaceous sandstones. This groundwater is recharged very slowly; the most recent water here is at least 20,000 years old. The mineralization of this water varies greatly and the least saline aquifers (salinity less than 5 g/L) are found predominantly in the Syr Darya basin. Part of the deep groundwater is artesian, and salty water is brought to the surface through natural upsurges, as is the case in the Ustyurt plateau, a limestone plateau between the Aral and Caspian seas. The potential resources contained across the Aral basin are around 100 km^3, and these aquifers are primarily fed through infiltration from the lakes and rivers in the Amu Darya and Syr Darya basins and from seepage from irrigation networks.

Vast aquifer complexes also exist in deserts in the Southern hemisphere. There are deep groundwater reserves in the dry eastern piedmont of the Cordillera of the Andes, fed by rivers from the mountains. These extend toward the east under the cold winter steppes of the Argentinian part of the Patagonian desert. In southern Africa, the Kalahari (Thomas and Shaw 1991) occupies a basin in the ancient Paleo-Mesozoic basement, raised toward the east and the south on the highlands of the Veld and the Cape, and buried under about 400 m of highly permeable red sand that is detritic material from the Tertiary and Quaternary. In Botswana and eastern Namibia, there is no desert, strictly speaking, but only a landform populated by acacias and baobabs, a steppe to the southwest and a savannah to the northeast, with significant summer rains (between 650 and 150 mm). However, this region has no surface water due to the high rate of evapotranspiration (>2,000 mm/year) and a near-total infiltration of water into the sandy cover. All significant water resources are underwater and, consequently, are largely out of reach of the people who live there (Bushmen).

Indeed, the only hydrogeological complex in the Southern hemisphere that can be compared to those in the Northern Hemisphere is the Great Artesian Basin in Australia (Habermehl 1980). Spread out over 1,700,000 km^2 and with a thickness of up to 3,000 m, it holds a total reserve of around 20,000 billion m^3 of water with very low salinity (0.5 – 1 g/L) (Margat 1983). This basin is also a syncline structure, filled with continental sedimentary detritic material (quartzites sandstones) ranging from the Trias to Cretaceous and interrupted by marine clay sequences that divide the structure into multiple superposed aquifers. The recharge of the aquifers, estimated to be around 1,100 million m^3/year, occurs largely through rainfall (on average, less than 125 mm), infiltration from tropical floods in the tributaries of Lake Eyre and seepage from the lake itself. On the other hand, there are heavy losses from the aquifers (around 30 million m^3/year), as there are several springs feeding flows at short distance and subsurface flows, only a part of which is re-injected through infiltration.

1.6. References

Bertouille, H., J.-P. Coutard, F. Soleilhavoup, and J. Pellerin (1979). "Les galets fissurés. Étude de la fissuration des galets sahariens, comparaison avec des galets issus de diverses zones climatiques." In *Revue de Géomorphologie Dynamique* XXVIII (2), pp. 33–48.

Birot, P. and F. Joly (1952). *Observations sur les glacis d'érosion et les reliefs granitiques au Maroc.* Vol. 3. Mémoires et Documents. Paris: Centre national de la recherche scientifique, pp. 9–60.

Bourrié, G. and P. Freytet (2008). "Résultats de la météorisation qualitatifs et quantitatifs." In. *LES FORMATIONS SUPERFICIELLES, Genèse - Typologie — Classification — Paysages et environnement - Ressources et risques*. Ed. by Y. Dewolf and G. Bourrié. Paris: Ellipses. Chap. 2.4, pp. 78–88.

Bourrié, G., F. Trolard, and P. Freytet (2008). "Altération chimique." In. *LES FORMATIONS SUPERFICIELLES, Genèse - Typologie — Classification — Paysages et environnement - Ressources et risques*. Ed. by Y. Dewolf and G. Bourrié. Paris: Ellipses. Chap. 2.4, pp. 52–73.

Braquaval, R. (1957). *Études d'écoulement en régime désertique - Massif de l'Ennedi et région nord de Mortchai*. Paris: ORSTOM. 92 pp.

Bryan, R. B. and S.-H. Luk (1981). "Laboratory experiments on the variation of soil erosion under simulated rainfall." In *Geoderma* 26 (4), pp. 245–265.

Capot-Rey, R. (1953). *Le Sahara français*. Paris: Presses Universitaires de France. 564 pp.

Carroll, D. (1970). *Rock weathering*. New York: Plenum Press. 203 pp.

Castany, G. (1967). *Traité pratique dee eaux souterraines*. 2nd ed. Paris: Dunod. 661 pp.

— (1982). *Principes et méthodes de l'hydrogéologie*. Paris: Dunod. 236 pp.

Cooke, R. (1981). "Salt weathering in deserts." In *Proceedings of Geologists Association* 92, pp. 1–16.

Cornet, A. (1952). "Essai sur l'hydrogéologie du Grand Erg Occidental et des régions limitrophes - Les foggaras." In *Travaux de l'Institut de recherches sahariennes* 8, pp. 71–122.

Cosandey, C. and M. Robinson (2000). *Hydrologie continentale*. Paris: A. Colin. 360 pp.

Coutard, J.-P. (2008). "Thermoclastie." In. *LES FORMATIONS SUPERFICIELLES, Genèse - Typologie — Classification — Paysages et environnement - Ressources et risques*. Ed. by Y. Dewolf and G. Bourrié. Paris: Ellipses, pp. 13–16.

Coutard, J.-P. and A. Journaux (1976). "Les expériences de thermoclastie." In *Bulletin du Centre de Géomorphologie du CNRS, Caen* (21), pp. 29–32.

Dewolf, Y. and G. Bourrié, eds. (2008). *LES FORMATIONS SUPERFICIELLES, Genèse - Typologie — Classification — Paysages et environnement - Ressources et risques*. Paris: Ellipses. 815 pp.

Dorn, R. I. (1984). "Cause and implications of rock varnish microchemical laminations." In *Nature* 310 (30 August 1984), pp. 767–770.

Dorn, R. I. and M. J. Deniro (1985). "Stable carbon isotope ratios of rock varnish organic matter: a new paleoenvironmental indicator." In *Science* 227 (4693), pp. 1472–1474. DOI: 10.1126/science.227.4693.1472.

Dorn, R. I. and T. M. Oberlander (1981). "Microbial origin of desert varnish." In *Science* 213.4513, pp. 1245–1247.

Dorn, R. I., D. Bamforth, et al. (1986). "Cation-ratio and accelerator-radiocarbon dating of rock varnish on Mojave artifacts and landforms." In *Science* 231 (4740), pp. 830–833. DOI: 10.1126/science.231.4740.830.

Dubief, J. (1963). “Contribution au problème des changements de climat survenus au cours de la période couverte par les observations météorologiques faites dans le nord de l’Afrique.” In. *Changes of Climate*. Arid Zone Research Series 20. Paris: UNESCO, pp. 75–79.

— (1953). *Essai sur l’hydrologie superficielle au Sahara*. Algiers. 457 pp.

— (1959–1963). *Le climat du Sahara*. 2 vols. Mémoire hors série. Algiers: Institut de recherches sahariennes. 312 & 275.

Duchaufour, P. (2004). *Introduction à la science du sol. Sol, végétation, environnement*. 6th ed. Paris: Dunod. 352 pp.

Evans, I. (1970). “Salt crystallization and rock weathering : a review.” In *Revue de Géomorphologie Dynamique* 19, pp. 153–157.

Feodoroff, A. (1965). “Mécanisme de l’érosion par la pluie.” In *Revue de Géographie Physique et de Géologie Dynamique* nouvelle série, VII, pp. 149–163.

Freytet, P. (2008a). “Altération biochimique.” In. *LES FORMATIONS SUPERFICIELLES, Genèse - Typologie — Classification — Paysages et environnement - Ressources et risques*. Ed. by Y. Dewolf and G. Bourrié. Paris: Ellipses. Chap. 2.5, pp. 74–78.

— (2008b). “Fracturation initiale des matériaux alimentant les formations superficielles: notions de tectonique et de mécanique des roches.” In. *LES FORMATIONS SUPERFICIELLES, Genèse - Typologie — Classification — Paysages et environnement - Ressources et risques*. Ed. by Y. Dewolf and G. Bourrié. Paris: Ellipses. Chap. 1, pp. 3–12.

Freytet, P. and E. Verrecchia (2008). “Induration et consolidation par les carbonates (tufs, travertins, croûtes calcaires).” In. *LES FORMATIONS SUPERFICIELLES, Genèse - Typologie — Classification — Paysages et environnement - Ressources et risques*. Ed. by Y. Dewolf and G. Bourrié. Paris: Ellipses. Chap. 8.1, pp. 337–350.

Gale, S. J. and S. Bainbridge (1994). “Megafloods in inland eastern Australia, April 1990.” In *Zeitschrift fur Geomorphologie* 38.1, pp. 1–11.

Godard, A. and G. Bourrié (2008). “Désagrégation granulaire, arénisation.” In. *LES FORMATIONS SUPERFICIELLES, Genèse - Typologie — Classification — Paysages et environnement - Ressources et risques*. Ed. by Y. Dewolf and G. Bourrié. Paris: Ellipses, pp. 43–51.

Goudie, A. S. (1989). “Weathering processes.” In Thomas, D. *Arid Zone Geomorphology*. London: Belhaven Press, pp. 11–24.

Goudie, A.S., R. Cooke, and I. Evans (1970). “Experimental investigation of rock weathering by salts.” In *Area* 2.4, pp. 42–48.

Habermehl, M. (1980). “The Great Artesian Basin - Australia.” In *Journal of Australian Geology and Geophysics* 5, pp. 9–38.

Horton, R. E. (1933). “The role of infiltration in the hydrologic cycle.” In *Eos, Transactions American Geophysical Union* 14.1, pp. 446–460.

Jansson, M. B. (1982). *Land erosion by water in different climates*. 57. 151 pp.

Joly, F. (1953a). “Essai d’une définition du Présahara marocain.” In *Bulletin de la Société Royale de Géographie d’Égypte* 25, pp. 57–70.

Joly, F. (1953b). "Quelques phénomènes d'écoulement sur la bordure du Sahara dans les confins algéro-marocains et leurs conséquences morphologiques." In *Comptes-rendus du Congrès International de Géologie* (Algiers, 1952). Vol. section VII, pp. 135–143.

— (1962). *Études sur le relief du sud-est marocain*. Travaux de l'Institut scientifique chérifien, série Géographie physique 10. Rabat. 578 pp.

— (1968). "Hydrographie aride." In. *Mélanges offerts à Maurice Pardé*. Grenoble: Ophrys, pp. 323–326.

Joly, F. and J. Margat (1955). "L'évolution récente du réseau hydrographique dans la plaine du Tafilalt (Sud-Est marocain)." In *Notes et Mémoires du Service Géologique du Maroc* 123, pp. 59–72.

Laudermilk, J. (1931). "On the origin of desert varnish." In *American Journal of Science* 21, pp. 51–66.

Létolle, R. and M. Mainguet (1993). *Les grands lacs d'Asie - L'Aral*. Paris: Science. 357 pp.

Loup, J. (1974). *Les eaux terrestres: hydrologie continentale*. Masson. 176 pp.

Margat, J. (1983). "Les ressources en eau souterraines sont-elles "minables" ?" In *Géochronique* (3), pp. 11–12.

Margat, J. and R. Saad (1984). "Les nappes souterraines profondes: des mines d'eau sous le désert." In *Nature et Ressources, Unesco* XX (2), pp. 7–13.

Millot, G. (1964). *Géologie des argiles*. Paris: Masson. 500 pp.

Mutin, G. (2000). *L'eau dans le monde arabe*. Paris: Ellipses. 156 pp.

Muxart, T. and P. Birot (1977). *L'altération météorique des roches*. Publications du Département de Géographie 4. Paris: Université de Paris IV, Paris Sorbonne. 279 pp.

Nicod, J. (2000). "Ras-el-Aïn (Syrie N) et le problème des sources karstiques artésiennes aux marges et à l'intérieur des domaines arides / (Ras-el-Ain (N Syria) and the problem of the artesian karst springs at the border or inside of the arid areas)." In *Bulletin de l'Association de géographes français* 77 (2), pp. 189–199. DOI: https://doi.org/10.3406/bagf.2000.2163.

Oberlander, T. M. (1994). "Rock varnish in deserts." In. *Geomorphology of desert environments*. Springer, pp. 106–119.

OSS (1995). *Ressources en eau des pays de l'OSS, évaluation, utilisation et gestion - Carte et notice*. Paris: Observatoire du Sahara et du Sahel-UNESCO. 80 pp.

Ozouf, J.-C. (2008). "Haloclastie." In. *LES FORMATIONS SUPERFICIELLES, Genèse - Typologie — Classification — Paysages et environnement - Ressources et risques*. Ed. by Y. Dewolf and G. Bourrié. Paris: Ellipses, pp. 26–29.

Peel, R. F. (1974). "Insolation weathering : some measurements of diurnal temperature changes in exposed rocks in the Tibesti region, central Sahara." In *Zeitschrift für Geomorphologie, Neues F.* Supplement Band, 21, pp. 19–28.

Peel, R. F., M. Chisholm, and P. Haggett, eds. (1975). *Processes in physical and human geography: Bristol essays*. London: Heinemann Educational. 417 pp.

Pissart, A. and J.-P. Lautridou (1984). "Variations de longueur de cylindres de pierre de Caen (calcaire bathonien) sous l'effet de séchage et d'humidification." In *Zeitschrift fur Geomorphologie* 49, pp. 111–116.

Prick, A. (1999). *Étude de la cryoclastie et de l'haloclastie par méthode dilatométrique*. Vol. XIX. Mémoire de la Classe des Sciences, 3e série. Louvain-la-Neuve: Académie royale de Belgique, Kevin Hall. 311 pp.

Riser, J. (2000). "Modelés karstiques, circulation souterraine et alimentation des lacs à sédimentation carbonatée du bassin de Toudmi, Sahara malien." In *Bulletin de l'Association de géographes français* (2), pp. 200–206.

Rodier, J. (1982). "La transposition des résultats des bassins représentatifs et ses problèmes." In *Cahiers de l'ORSTOM, série Hydrologie* 19.2, pp. 115–127.

Roose, É. (1977). *Érosion et ruissellement en Afrique de l'Ouest*. Travaux et Documents 78. Paris: Éditions de l'ORSTOM. 108 pp.

Salomon, J.-N. (1987). *Le Sud-Ouest de Madagascar : étude de géographie physique*. 2 vols. Bordeaux: Presses de l'Université de Bordeaux. 1008 pp.

— (2000). *Précis de Karstologie*. Collection Scieteren. Bordeaux: Presses de l'Université de Bordeaux. 250 pp.

Sanlaville, P. (2000). *Le Moyen-Orient arabe. Le milieu et l'Homme*. Collection U. Paris: Armand Colin. 264 pp.

Savat, J. and J. De Ploey (1982). "Sheetwash and rill development by surface flow." In. *Badland geomorphology and piping*. Ed. by R. Bryan and A. Yair. Geo Books: Norwich, pp. 113–126.

Schick, A. P. (1970). "Desert floods. Interim results of observations in the Nahal Yael research watershed, Southern Israel, 1965-1970." In *IAHS-UNESCO Symposium: Representative and Experimental Basins, Wellington, NZ*, pp. 478–493.

Schick, A. P. and J. Lekach (1987). "A high magnitude flood in the Sinai desert." In *Catastrophic flooding* 18, pp. 381–410.

Scoging, H. (1989). "Runoff generation and sediment mobilisation by water." In Thomas, D. *Arid Zone Geomorphology*. London: Belhaven Press, pp. 87–116.

Smith, B. J. (2009). "Weathering processes and forms." In. *Geomorphology of Desert Environments*. Ed. by A. J. Parsons and A. D. Abraham. 2nd ed. Amsterdam: Springer Netherlands, pp. 69–100.

Soleilhavoup, F. (1977). "Les cailloux fissurés des regs sahariens : étude descriptive et typologie." In *Géologie Méditerranéenne* IV (4), pp. 335–364.

Thomas, D. (1989). *Arid Zone Geomorphology*. London: Belhaven Press.

Thomas, D. and P. Shaw (1991). "Relict'desert dune systems: interpretations and problems." In *Journal of Arid Environments* 20.1, pp. 1–14.

Tricart, J. (1960). "Manifestations périglaciaires du Quaternaire ancien en France." In *Biuletyn Peryglacjalny* 9, p. 143.

— (1981). *Précis de Géomorphologie, 3 - Géomorphologie climatique*. Paris: CDU-SEDES. 313 pp.

Vanney, J. R. (1960). "Pluie et crue dans le Sahara Nord-occidental." In *Travaux de l'Institut de Recherches Sahariennes*. Monographies régionales 4, p. 114.

Walther, J. (1893). *Allgemeine Meereskunde*. 195. Quelle & Meyer.

Yaïr, A. and H. Lavee (1974). "A real contribution to runoff on scree slopes in an extreme arid environment - A simulation rainstorm experiment." In *Zeitschrift für Geomorphologie* 21, pp. 106–121.

2

Temporary Water Bodies and Lakes

However paradoxical and surprising it may seem, a veritable "desert hydrology" (Joly 1968) does exist. A satellite image or a flight over a desert is enough to be convinced of the truth of this. One may observe temporary water bodies and flows created in recent rains, which leave hardly any trace in the soil, as well as permanent *lakes* and ancient *hydrographic networks*, etched into the topography during one of the "pluvials" in the distant (or recent!) past. These forms are testimonies of a hydrography that is largely inherited from the past and whose chief characteristic is the absence of any perennial flow at present and the consequent accentuated degradation in the existing networks (Dubief 1953; Joly 1962), as well as an almost-complete generalization of closed drainage systems or *endorrheic* systems. Only the large formations made up of dunes (ergs) and certain hyper-arid spaces are truly *arheic*, i.e. totally lacking any flow. On the other hand, at a regional level, there are also large *allogenous rivers*, which arise outside of the truly desertic region, but travel through the desert and sustain the life around their course.

2.1. Water bodies

The presence of stagnant water bodies is the result of a combination of favorable conditions that often occur in a desert environment. The chief of these, at a topographical level, is the existence of a number of closed, endorrheic depressions (Martonne and Aufrère 1928), that is, depressions bordered by relief features whose altitude is larger than the highest point reached by the water body. Some of these features can be born out of general geological conditions: craters caused by tectonic events, meteors or volcanic activity. From this point of view, the relief features are

---

Chapter written by Fernand JOLY.

not very different from those found in other regions of the world. Other features are more directly dependent on the geomorphological evolution of a region itself, whether present or inherited, i.e. the climate and its variations over time. Finally, other features can be born out of the disappearance of an organized hydrographic network, following the capture, depletion of flow, or drying out of the climate.

The climate of a region plays a principal role in the specific profiles of their water bodies, which make up so many individual hydrological units. The climate governs the quantity of precipitation received, which is the primary source of water coming in, and the rate of evaporation, which counteracts the precipitation. Climate also determines pluviometric contrasts between basins and surrounding heights, as well as the pattern of alternating humid and dry periods across all timescales. However, the hydrological balance for water bodies that must survive in these harsh conditions is generally more complex: in addition to the water coming in from rain, there is water brought in through inflows, diffuse run-off along the slopes and floods in the rivers (when they occur), upsurging and resurging of groundwater, and snow and ice in some cases. However, these are balanced by losses through infiltration, seepage and human activity.

The water bodies that are present in arid regions today can be broadly categorized into two types. The first are those that only have a precarious existence, where humans play a primordial role given their dispersion. The second are more localized, permanent (or at least multi-year) features of the landscape.

2.1.1. *Elementary water bodies*

The desert has a wide variety of water bodies: stand-alone, scattered, poorly supplied, but all highly important and much-sought after by those who live in deserts. These water bodies are most often small and intermittent, containing freshwater or saltwater, and are known to the nomads and travelers (albeit, under a variety of names) whom they sustain. Every downpour leaves a number of *pools* of water in hollows or in sheltered spots:

> *After the rains, not all the water escapes the mountains and get absorbed into the sands; almost everywhere, some water remains in the holes. Small pools will dry out by evening or the next day, mid-size ones will last for a week, others for a month, four months, six months or even longer, and the large pools, that is, the deep pools, will endure as long as needed till the next rains, for years.* Th. Monod, *Méharées*

> *"Water! Water lay everywhere; in pools both great and small; in depressions and ancient sand channels. Fading in memory was that dusty-grey-brown half-dead dehydrated land; [...] for here and now was*

the real world of sparkling waters, and [...] the vivid green of grass and herbage". A. Upfield, *The Torn Branch*

Most water bodies only last as long as it takes for the water to get absorbed into the soil. They last long enough, however, to allow the germination of seeds that have fallen which, at times, allow the desert to be covered in flowers or get greened. These form the *acheb* that Saharan pastoralists use, a true windfall for their herds.

Deep hollows or large depressions hold larger ponds for a longer time and these are also more valuable. In the Sahara, a residual and semipermanent pond that occupies a hollow in a rock, or at the bottom of a wadi in a valley, is called a *guelta* (or *glat*). These are a sort of natural reservoir, sometimes surrounded by some palm trees, whose positions are fixed and known. They also serve as drinking pools for wild, as well as domesticated, animals. Some of these ponds are simply fed by the rains, are residual waters from floods, or are temporary upsurges from underflows. These dry out easily. Other ponds are enduring and are maintained by seepage or springs (gravity or artesian springs). Some ponds are very old and may have preserved some unique relict fauna like shrimps, fish or dwarf crocodiles (the Archei gueltas in Ennedi, or the Matmata guelta in Tagant).

Larger ponds (from some ten meters to a few hundred meters in diameter) temporarily occupy closed depressions that are more or less circular and have a flat bottom. In the septentrional Sahara region, these depressions are called dayas. These are either true *dolines* (in limestones) or simple deflation basins (blowouts). The water bodies that form in these in the rainy season, or after a very heavy shower, are quite shallow (1–2 m deep) and normally contain freshwater or water with low salinity. The presence of these water bodies facilitates the chemical alteration of the floor of the depression, especially the decarbonatation of limestone. At the same time, the water bodies collect fine debris swept down the slopes by diffuse run-off streams. When dry, the bottom of these depressions is clayey and cracked, or clayey-silty and capable of supporting bushy vegetation that is more or less *halophytic*. Dayas are quite frequent in the septentrional Sahara (the plateau of Dayas, in eastern Algeria) and are also found in similar forms in South Africa (called *vleis* in Afrikaans), Australia (called *pans*) and Central Asia (*tsaidams*).

The groundwaters held within the large ergs sometimes flow out into hollows, either temporarily or permanently, in the form of small *interdune ponds*. The dried out bottoms of these ponds are often marked by a slight vitrification of the sand, or by a thin, hardened layer of clay or gypsum. Around the periphery of the ergs, at the interface between the dunes and their rocky substrates, the deepest water tables feed many springs, which are often quite large and sometimes spread out into proper small lakes that attract plant and human life.

2.1.2. *Playas*

In the deserts of Mexico and the Western United States, the alluvial fan floodplains that extend the debris-covered lower slopes of mountains (*bajada*, or *bahada*) at the center of a vast endorrheic basin (*bolson*), are called *playas*. Stretching over a kilometer or longer, these are almost equivalent to the North African *chotts* or the Mongol *gobis*. The playa is the topographical and hydrological base of the bolson. It is the terminal point of all the slopes and surface flows, especially tributaries. It also captures waters from groundwater tables in the slopes. On the other hand, however, a playa is also the site of significant water loss, not only through evaporation (which is very active on stagnant and shallow water), but also through infiltration, depending on the soil's infiltrability and permeability.

The magnitude of the floods depends, of course, on the size of the watershed, the nature of the substrate and climatic conditions. The smaller playas are the most inconstant. In the absence of an adequate catchment area, they are more often filled by local showers, rather than by regional rains. They lose more water through evaporation than through infiltration; they dry out faster than larger playas and most of them are ephemeral. Larger playas are more dependent on seasonal patterns and ground water discharge. Some playas have significant salinity, depending on the drainage from the slopes, the material brought in through inflows and by the wind and, above all, on concentration by evaporation. Playas with a high salt content are called *sabkhas* in the Sahara, *kavir* in Iran, *takyrs* in Turkestan, *nors* in Mongolia, *salt pans* in the United States, South Africa and Australia, and *salinas* or *salares* in South America and Mexico.

Most playas are subject to an alternation between short, wet periods and longer dry periods, which frequently last over a year. As a result of this, the water levels in playas are variable, and as the banks are low and flat, variations in volume result in a considerable shift in the banks and unique morphological and biological zonation (Tricart 1954; Mabbutt 1977). Subsurface limestone crusts (calcretes) or gypsum crusts form on foreshores that undergo short submersion periods and long dry periods, and can extend on several kilometers. These crusts are partly linked to intense underground discharges, when the piezometric level of the water rises through very permeable sediments. These foreshores are generally bare or covered by halophytic, steppic plant associations that tolerate a temporary submersion, comprising of Chenopodiaceae and Salsolaceae[1]. This vegetation offers valuable grazing and makes up the *chotts*, strictly speaking. When the season allows, this land is also used for the occasional and random cultivation of cereals (the *grara*, as the basin is called in the western Saharan region). In sabkhas, the proper salts, bare of all

1 Now included in Amaranthaceae.

vegetation, occupy the center of the playa, in the form of powdery deposits called *fech-fech* (Volume 3, Chapter 1, Figure 1.9), with friable inflorescence or, sometimes, crystal twinning, seen as "spearhead" shapes or the gypsum desert rose, mixed with granular silts and clays.

When water disappears in the dry season, the entire playa is subject to wind deflation. In the denuded zones, efficient winds lift up and carry small grains and aggregates (<2 mm) deposited by floods or generated in the wet soil when water was high in the basin. Selective deflation among granular or silty-clayey sediments sometimes leads to the formation of parallel furrows and crests formed in the direction of the prevailing wind. These are called *yardangs* (Volume 3, Chapter 1, Figure 1.8) in Turkestan, *kaluts* in the Iranian Lut (Dresch 1968; Dresch 1975) and *sand ridges* in Australia. These are often extended downwind by lines of longitudinal dunes. In regions richer in clay and salts, the surface develops a fragmented appearance, seemingly tiled with polygonal, mosaic-like formations, delineated by cracks or ridges measuring a few decimeters, formed by the extrusion of the underlying salty mud, and more or less smoothed down by the floods. Part of this debris is carried away by the prevailing wind and sticks to the vegetation along the periphery of the sebkha, facing the wind. This often results in a concave and asymmetric wall being built up through accumulation, with a steep slope on the windward side. This is formed of silt mixed with clay and/or salt, and is stabilized through compaction or hygroscopic consolidation into a sort of parabolic dune, called a lunette (Volume 3, Chapter 1, Figure 1.36, (Hills 1940)). The rest of the debris is spread out across the area of the bolson and even beyond that.

Comparable phenomena can be observed along the course of rivers, in alluvial plains that have built up on either side of a river breaking through hard rock (*foum*) or flowing through a gorge carved into hard rock (*kheneg*). The sharp narrowing of the riverbed upstream leads to an accumulation of the solid load carried by the water and the creation of a floodplain. Downstream, at the outlet of the narrowed neck, sheet floods occur and can build up a very flat dejection cone, called an *alluvial fan*, which can end up in a playa. However, the difference between these fans and playas is that the water and land in the fans are less salty, or have no salinity at all. They are also suitable for intermittent agriculture (constituting the *mader* found in the Moroccan Sahara). These fans are also subject to intense deflation in dry periods.

2.2. Lakes

Like underground aquifers, *lakes* (Pourriot and Meybeck 1995) are a very valuable resource in deserts. They form potential water reserves on the surface that are relatively stable (at least on a human scale), although they are fragile and difficult to use, and are, in all circumstances, exceedingly rare. These lakes vary in size (ranging from a few square kilometers to several thousand square kilometers), and depth (ranging from a few meters to a few hundred meters deep) and are proper water traps integrated into local hydrographic systems (Touchart 1996). While the greatest number, and the largest of these lakes, are found in Asia (Létolle and Touchard 1998), lakes exist in all deserts in the world, whether in Africa, the Americas or Australia.

These water bodies can be divided into various categories, listed below with examples, with a broad division between *inland seas* (the last surviving proof of how far inland the oceans had extended in earlier ages, or a Sound, born of a sea that no longer exists) and *endorrheic lakes* (closed basins with no surface drainage or outlet).

2.2.1. *Inland seas*

The most remarkable examples of these water bodies are the great lakes found in the Aral-Caspian or Turanian (Turkestan) depression. Up to the Tertiary period, a large oceanic corridor existed between the northern part of the Eurasian continental belt and the southern part of the Afro-Arabian plate that was moving northeast. This was the Tethys Sea, from which the Iranian and Caucasus ranges emerged during the Mio-Pliocene. The formation of these mountains isolated the Paratethys or "Sarmatian sea" to the north, which gradually retreated into distinct basins and grew less maritime, separated from the ocean by raised banks. The Turanian depression is the easternmost of these basins and has an opening to the north, while in the south, it abuts multiple mountain ranges: the Caucasus, the Elburz, the Hindu-Kush and the Pamir. The only surviving connection with the Mediterranean today is the "Ponto-Caspian basin", a narrow corridor between the Caspian Sea and the Sea of Azov, a remnant of the late Tertiary period, today traversed by the East and West Manych rivers. The only remnants of the former Tethys Sea in this region are the two large lakes that still bear the title "Sea": the Caspian Sea to the west and the Aral Sea to the east, separated by the Ustyurt plateau fault blocks.

2.2.1.1. *The Caspian: Central Asia.* 42°0′*N*, 50°30′*E*

This is the largest lake in the world and the oldest lake known to western geographers and merchants (Létolle and Touchard 1998).

> *"[...] the Caspian is separate [from the Black Sea or Pont Euxin]. Its length is what a ship rowed by oars can traverse in fifteen days, and its breadth, where it is broadest, is an eight days' journey. Along its sunset shore, stretches the range of Caucasus, which has more and higher mountains than any other range [...]; towards the east and the sunrise there stretches from its shores a boundless plain as far as sight can reach"* (Herodotus ca. 443 BCE–420 BCE, vol. I, pp. 203–204).

In the 3rd Century BCE, Eratosthenes thought that the Caspian or Hyrcanian Sea was a deep gulf off the Northern Ocean, which bordered the Ekoumène to the North. Strabo used the diagram in the 1st century CE. It was not till the 2nd Century CE and Ptolemy's work, that the Caspian was represented on a map as a lake. However, the idea of the gulf endured for a long time. In Medieval times, travelers from Western Europe went around the Caspian via its northern shores and finally did it justice. For example, we have William de Rubruck's 13th Century account by (Rubrouck 1255, chapter IX):

> *So we came to the Volga, the greatest of rivers [...] four times larger than the Seine [...], emptying into a certain lake, or sea, which is now called Sea of Sirsan, from a certain city on its coast in Persia. Isidorus [of Seville], however, calls it the Caspian sea, for it has the Caspian mountains and Persia to the south [...] and to the north is this wilderness in which are now the Tartars [...] And on that side it receives the Volga, which rises in summer as does the Nile of Egypt [...] this sea has mountains on three sides, but on the north there is only a plain [...].One can go around it in four months, and it is not true, as stated by Isidorus, that it is a gulf of the Ocean. It nowhere reaches the Ocean, but is everywhere surrounded by land.*

The "Sea", which extends across 390,000 km^2, is situated 23 m below sea level and its average depth is 330 m. The deepest points are found in the southwest, divided into two basins by the Apcheron Peninsula. The southern basin, bordered by the Elburz Mountains, is the deeper basin, with its depth exceeding 1,000 m, while the northern basin has a depth of less than 900 m. The entire northern and northeast part of the lake is a platform located at a depth of less than 100 m, often less than 15 m deep. The southern shores at the foot of the Elburz Ranges and around the Apcheron Peninsula are quite steep, except around the mouth of the small coastal rivers that empty into the lake. All the other shores are flat, or are bordered by ancient terraces. There are, however, two notable exceptions: the first, to the north, is the delta of the Volga, a large, humid, alluvial zone and a biological reserve measuring around 600 km^2; the second, on the coast, is the Kara-Bogaz-Gol Gulf, which has its own hydrology and sedimentology. While not being an entirely desert environment, the Caspian is surrounded, at least to the north and the east, by frankly dry territories: less than 200 mm of summer rainfall in the north, and winter rainfall in the south, with a dry period lasting six or seven months. The lake is located at the eastern end of the large, arid corridor (cold desert: −7°C in winter, 25°C in summer) in Central Asia, and the climate this engenders is a major factor in the life of this sea.

Apart from the rain and a few small coastal rivers, the Caspian Sea is also fed by two significant inflows, both located on its northern shores: the Ural and the Volga.

The Volga, which is the most powerful of its inflows (8,000 m^3/s), is also the single largest source of water to the Caspian, making up 95% of its water supply. This water is cold and is either freshwater or mildly brackish, following the river's journey over the salty steppe soil in the lower valley. The water flowing in from the Volga sets up a circular surface current that descends all along the western shore and rises along the eastern shore, with accompanying whirlpools forming above underwater trenches. Infiltration seepage feeds the underwater reserves, while the largest losses are due to evaporation (over 1,000 mm/year), which is stronger in the warmer south than in the north, where it is mitigated by the ice that forms over the sea from December to April. Salinity is highly variable, but is lower in the north (10 g/L) because of the freshwater brought in by the rivers, especially in late spring when the snow and ice melt. The salinity in the south (13 g/L) results almost entirely from evaporation. This salinity is also responsible for the unique nature of the Kara-Bogaz-Gol gulf. This is a lagoon separated from the sea by a rocky ridge, but connected through a canal and surmounted by a sandy spit. The lagoon is situated 4 m below the Caspian and is found in a true desert. The water that comes into the lagoon leaves only through evaporation. When the natural canal connecting it to the Caspian sea was blocked off between 1980 and 1985, the gulf dried out entirely. This resulted in even higher salinity (as high as 290 g/L), which increases as we move further away from the Caspian Sea. Chiefly gypseous ($CaSO_4 \cdot 2\,H_2O$), this sedimentation is enriched with salt (NaCl) toward the northeast, which is actively quarried for salt.

The Caspian Sea is a permanent lake. Its volume (around 67,000 km^3) is too large for it to completely dry up over a human timescale (or an even longer historical timeframe). On the other hand, while it has always remained below sea level, it has constantly varied over time, as can be seen from sedimentological, geomorphic, and archeological evidence. Short-term variations, seasonal variations (water brought in through floods in the rivers in summer, ice retention in winter) or occasional occurrences (a rise in the water level in the west following a flood current along the coast, or a strong wind from the northeast) only produce a variation of tens of centimeters. Variations over longer periods are larger: there were variations up to a few hundred meters during the Quaternary, and 20–40 m over 10,000 years. In the early 19th Century, as in the late Medieval period, the water level was very high (around −22 m). From the 19th Century onwards, it constantly decreased, with a few small oscillations, until 1937, when it reached its lowest point of −29,04 m. From 1937 onwards, it once again started rising at the rate of tens of centimeters per year until 1995, when the movement seemed to stop, if not regress. There could be several causes for this, but at any rate in the past few decades there has been a slow transgression onto the surrounding flat shores, resulting in the water table rising and the consequent salinization of the soil, and progressive disappearance of the beach, the spits and marginal lagoons, significantly modifying the whole physical coastal landscape.

2.2.1.2. *The Aral Sea: Central Asia* 45°0′*N*, 60°0′*E*

Like the Caspian Sea, the Aral "Sea", or lake (Létolle and Mainguet 1993) is the only surviving part of the Sarmatian sea. While it has been separated from its neighbor since the Pliocene, it still retained a connection with it until the 17th Century, through a now-extinct water body (the Uzboi). This water body was channeled through a paleovalley, that is today obstructed by sediments from the Amu Darya delta. In 1960, when it extended over a large area, its altitude was 53 m, its volume was 1,080 km^3, and its surface area was 66,000 km^2, about one-sixth of the Caspian. The lake is divided into two highly unequal basins, which are separated today by a raised ridge. To the north lies the Lesser Sea or Smaller Aral Sea, the bottom of which lies 22 m below the benchmark level of 53 m; in the south is the Greater Sea or Larger Aral Sea[2], whose center is occupied by an underwater plain at a depth of 25–30 m, while beyond a chain of islands, a tectonic "trench" runs along the western coast, going as deep as −68 m, or 15 m below sea level. The Aral is the lowest point of a drainage basin that extends over 1,500,000 km^2, enclosed to the north by the hills of Kazakhstan and to the southeast by some of the highest mountain ranges in the world (over 5,000 and 7,000 m): the Hindu Kush Ranges, the Pamir, and the Tian Shan. The lake itself is entirely enclosed within a desert and, on its western shores, is bordered by a cliff stretching to a height of 150–200 m. In earlier times, these cliffs were battered by storms and carved out of the fluvial-marine sediments of the Cretaceous and Tertiary periods, which offered little resistance to the elements. The other shores are low and not very clearly delineated. They are quite close to the Kyzylkum dunes and are only indented by the Syr Darya and Amu Darya deltas, the first in the northeast and the second in the south. These are marshy areas divided by several streams from the rivers and ringed by bushy vegetation.

The hydrological profile of the Aral Sea is that of a lake in a zone of cold desert aridity, with harsh winters and hot summers. In summer, the temperature of the surface of the water can go up to between 24 and 27°C, while in winter, the water freezes (the freeze setting in from around −1°C, in this salty water) for four to five months. One of the consequences of this freeze, which is also seen in the Caspian, is that the potential evaporation, which is high throughout the year (over 1,000 mm) is somewhat mitigated in the winter. With an average precipitation of 100–150 mm/year, the water balance remains largely negative and, indeed, if not for the water brought in by two large allogenous rivers (the Amu Darya and Syr Darya) the Aral Sea would have disappeared long ago. That being said, after a difficult journey through the desert, the two rivers supply very little water to the Aral Sea. Before 1960, the Syr Darya was only just able to maintain the level of the Lesser Sea,

2 Locally, in Russian, the Lesser and Greater Seas are referred to as the Maloye More and the Bolshoi More, respectively (translator's note).

while the Greater Sea only largely received water through precipitation and a residual flow from one of the branches of the Amu Darya. In total, the average inflow toward the Aral was only between 54 and 66 km^3/year, around two-thirds of which was brought in by the Amu Darya (Létolle and Mainguet 1993; Létolle and Touchard 1998). This was augmented by 6 km^3/year of precipitation and 2 km^3/year of exchange with the shores and the bottom. In such conditions, the hydrological balance was extremely fragile and, like the Caspian, the Aral Sea constantly underwent significant changes in its level. Older beaches have been identified between the absolute altitudes 30–58 m, and Quaternary layers, perhaps of tectonic origin, have been identified (because of fossil shells) up to depths of 60–65 m. Conversely, the current emersion has revealed traces of river terraces submerged long ago, evidence of older periods when the basin dried up (15th–16th Centuries CE). Until 1960, however, it was still possible to speak of stability, despite the diversion of the tributaries for irrigation since Antiquity, intensifying in the late 19th Century following Russian colonization. Seiches and variations caused by the contribution from the rivers continue to affect the level of the water bodies, causing changes of a few decimeters. Given how flat the shores are, these changes are enough to allow large-scale transgressions onto the shore. However, since 1960, the Aral Sea has seen a constant fall in its level, with an increase in systems that divert water from the rivers and with the droughts between 1970 and 1980. Inflows were almost completely absent and the water that did come in was largely through drainage from irrigated crops upstream. Between 1960 and 2003, the Aral Sea shrank from 66,000 to 22,000 km^2, its volume reduced from 1,080 to 160 km^3, and its level dropped from 53 to 31 m (Létolle and Mainguet 2004).

This shrinking resulted in a radical change in the physiography and hydrology of the lake. The Lesser Sea, blocked by a sandy isthmus, is now only connected to the Greater Sea through an artificial canal, but continues to be fed (poorly) by the residual water of the Syr Darya, and is drying at a slow rate. The Greater Sea, deprived of water from both rivers, dried out at a much faster rate and continues, inexorably, to dry out. The large lake that was seen in the 1960s is now practically divided into two parts: a central pool, a few meters deep, and a coastal basin to the west, which is a little deeper (around 50 m), separated by a meridian headland that connects former islands. The shores moved everywhere, by a few kilometers to the west, and up to 20 or 30 km in the south and the east. Aralsk, a port in the northeast that was once very active, is now located over 100 km from the “Sea”. The surface that has emerged, clayey-silty-sandy, is continuously affected by shrinkage cracks from the desiccation. The carcasses of abandoned ships jut out here and there in this saline plain, subject to heavy deflation. In the ever-shrinking waters and the pools from floods, salinity varies (from an average of 10 g/L to over 45 g/L) and calcium carbonate, gypsum and salt crusts have formed, more or less brittle, swept up by the wind. Halophytic vegetation (Chenopodiaceae, Salicornia and even Saxaul) try, with some difficulty, to colonize this terrain. The Aral Sea is dying and no one seems to know quite how to revive it – or if it is even worth the effort.

Saving the Aral Sea in its entirety seems an impossible task. This realization led to the idea of artificially enlarging the Lesser Sea, by erecting a dam on the isthmus. The first earth bank, constructed between 1992 and 1996, made it possible to raise the level of the Lesser Sea by at least 1.5 m with respect to the level of the Greater Sea. However, at the same time there was accelerated sandy-clay sedimentation and salinity increased dangerously as a result of evaporation and a lack of freshwater. This structure then broke up in 1998, revealing its fragility, and plans for its reconstruction brought up several political problems between Kazakhstan to the north, and Uzbekistan to the south, with the latter wishing to preserve some of the water brought in by the Syr Darya. However, despite the Uzbeks' protests, the World Bank chose to sacrifice the Greater Sea to the south and prioritize the Lesser Sea to the north. The World Bank financed the new construction on the isthmus: a sloping sandy dyke built from crushed stone and concrete that is submersible and fitted with sluices that can be opened to release excess water from the Lesser Sea. At the same time, the Syr Darya spillway was improved to maintain the area of the lake at around 5,600 km^2 and maintain its volume at around 72 km^3, to counteract the effects of evaporation, limit desertification, and revive navigation and possibly fishing as well. The Greater Sea, on the other hand, ended up dying out. Today it only receives around 10 km^3/year of water, chiefly salty water, carrying a heavy load of alluvions, from the Amu Darya. The water level continues to decrease and salinity continues to increase, being as high as 75 or 80 g/L in the western basin. The central basin is now a lifeless sabkha whose salinity touches 150 g/L.

2.2.2. *Endorrheic lakes*

Like playas, endorrheic lakes are the bottom of a hydrographic basin that may be very large (Lake Chad, Lake Eyre) and the final remnants of an isolated "hydroclimatic" system (Frecaut 1979). Consequently, their characteristic features (hydrological balance, sedimentology, chemical composition and variability) depend on the arid conditions prevailing in the region considered.

2.2.2.1. *Subtropical desert aridity*

2.2.2.1.2. The Dead Sea, Near East 31°30′N, 35°30′E

The endorrheic complex that makes up the Dead Sea consists of two large water bodies: Lake Tiberias and the Dead Sea itself. These are located at two ends of the Ghor lowlands and are connected by the Jordan river (Sanlaville 2000). This corridor is a tectonic fault, the Dead Sea Rift, running north to south, a fracture in the African continental plate caused by an Arabian subplate progressively moving away to the northeast, from the Neogene period onwards. This is a major fault line in the structure of the Near East, lying along the axis of the Lebanese Beqaa Valley, and extending south, above an altitude of 22 m, through the Gulf of Aqaba and then the Red Sea Rift, as far as the Horn of Africa. At this point, in a closed and arid environment, we find a variety of accumulated continental deposits: tectonic breccias, lava flows, lacustrine gypseous marl and fluviatile conglomerates. Topographically, the rift is a deep depression, well below sea level (−740 m in the middle of the Dead Sea), narrow (between 5 and 25 km) and depressed (on average, 1,000 m subsidence) between the marly-limestone plateaus (that grow increasingly arid to the south) of Galilee, Palestine and Israel to the west, and Syria and Jordan to the east.

Isolated from the sea, the Ghor receives all the water in the vicinity: water from groundwater tables and karst aquifers; water from small wadis on the slopes; and, most importantly, water from the Jordan river, which is the drainage basin for the entire hydrographic system of the region. The Jordan arises at the point of convergence of several upsurges on the western flank of Mount Hermon (2,814 m), on the southern tip of the Anti-Lebanon Mountains. Guided by tectonics, the upper course of the Jordan extends south of the Lebanese Litani river valley, which is deviated to the Mediterranean. After crossing the Hula Valley (altitude of 70 m), a marshy region dotted with freshwater or brackish lakes, the Jordan travels through gorges formed by Pleistocene volcanic activity before arriving at Lake Tiberias. This lake (Mutin 2000), the "Sea of Galilee" in which Christ's disciples went fishing, is a water body spanning 160 km^2 and is 45 m deep. It is blocked downstream by the dejection cone of the Yarmuk wadi and is located 210 m lower than the Syrian Golan. The Jordan discharges about 560 million m^3/year of freshwater into this lake, which spreads out across the surface. In addition, the lake receives notable inflows from sublacustrine springs (often salty), around 230 millions m^3/year. After exiting Lake Tiberias, the Jordan meanders on a winding course of 220 km, to eventually cover the 105 km between Lake Tiberias and the Dead Sea. This final stretch plays a key role in determining the state of the Dead Sea.

Some 500 million m^3 of brackish water is carried away from Lake Tiberias by the Jordan annually. En route to the Dead Sea, the river also has lateral inflows of freshwater from large tributaries such as the Yarmuk (500 million m^3/year), which comes through Syria, or from the Zarka (100 million m^3/year), which comes through Jordan. In total, the "natural" flow of the Jordan at the inlet into the Dead Sea, close to Jericho, can be around 1,450 million m^3/year. However, the water diverted for irrigation, pisciculture and, above all, to supply cities with water, removes almost 90% of water from the river. Consequently, the Dead Sea receives only a stream of waste water or drainage water, and even this can cease altogether in the summer. In addition, this part of the Jordan Valley has about 30 m of continental deposits from Ghor, the majority of which are gypseous marls and clays and quaternary saline deposits, which contribute to the suspended and dissolved load of the river. Although there is highly variable salinity here (ranging from 8 to 30 g/L) depending on the flow, the final stream from the Jordan river that flows into the Dead Sea is always saline.

During the Quaternary, the Dead Sea underwent significant changes in volume and level, due to climatic factors (Létolle and Touchard 1998; Sanlaville 2000). Subaerial morphological traces (dejection fans, terraces), now submerged, testify to the lower levels that the Sea touched during the last glaciations. Conversely, large water bodies occupied the site after the glaciers melted but then slowly shrank under the effects of aridity and also as a result of water being removed by humans. The most recent of these large lakes was Lake Lisan, which occupied the Ghor around 40,000 or 30,000 BP, during the large Wurm interglacial. Note that at 230 km long and located at an altitude of –180 m, Lake Lisan left behind marlstones and lacustrine gypseous deposits that are spread out as far as the shores of Lake Tiberias, forming the substrate and environment of the current "Sea" there. During the Holocene climatic optimum, around 9,000 BP, there was still an extant, if receding, lake in the middle of a Mediterranean forest. In the 1970s, this lake still existed as a water body measuring 136 km^3, with an area 940 km^2 (85 × 17 km), at an altitude of –397 m. From the 1970s onwards, the level decreased again by a dozen meters, touching –408 m or –410 m. In one century, the Dead Sea lost one-third of its area. In 2003, there was nothing but a much smaller lake in a desolate landscape with bare slopes, badlands, muddy beaches and salt deposits.

The Dead Sea has long attracted the attention of travelers and curious visitors, because of its significant history as well as its strange behavior and environment. Scientific studies on the nature of the Dead Sea can be traced back to the 18th Century, when Lavoisier studied the chemical composition of its waters. Research continued all through the 19th Century (Angelot 1843) and up to the present, especially in Israel (Neev and Emery 1967). Observers have been most struck by the water's density (1.16) and salt content (300 g/L), and all of them have mentioned the fact that it is hard to immerse a human body. The salinity results from a chemical amalgam chiefly composed of chlorides of magnesium ($MgCl_2$) and sodium ("sel" NaCl). However, there are also potassium salts (KCl), sulfates (gypsum $CaSO_4 \cdot 2H_2O$), carbonates and bromides. Nevertheless, the most characteristic feature is the marked stratification of the water in the depths. Surface waters, up to a depth of about 40 m, have an average salinity of 300 g/L, a high oxygen content and a temperature that varies from around 10°C in winter to 38°C in summer (sometimes going beyond 50°C in the border lagoons). This water is relatively 'light' water, which does not sink and recharges very slowly. This layer 'floats' on the deeper waters, which are denser. There is almost no vertical or lateral mixing, given the absence of a current. Beyond 40 m, the water becomes anaerobic, with no oxygen, and salinity increases to over 320 g/L, forming large salt deposits through saturation.

At the same time, the temperature homogenizes at around 22°C. These conditions are not favorable for life-forms.

The fact that magnesium chloride is more prevalent than sodium chloride, and the absence of iodine in the composition of the water in the Dead Sea has sometimes been used to argue that there could be no marine heritage for the water body. The salt in the Dead Sea is essentially washed in from the gypsiferous-saline continental and volcanic formations in the Ghor, as well as an ongoing concentration of the waters of the lake, resulting from evaporation. The local hydrological balance would seem to support this point of view. There is practically no fluviatile influx that comes from a great distance away, apart from the weak (and salty) contribution from the Jordan River. The annual direct pluviometry over the lake, following the Mediterranean pattern of maximum precipitation during winter, does not exceed 50–60 mm. The equivalent of 1,500 mm comes in from surface runoff on the Jordanian slopes and from a few winter showers over the Negev. In addition to this, around 250 mm of water comes from springs (which are themselves quite salty) around the lake as well as in its depths. There is considerable loss of water through infiltration (around 150 mm) and especially through evaporation, which accounts for about 1,600 mm (Létolle and Touchard 1998) or even 2,400 mm (Sanlaville 2000), creating a fragile and variable steady state that is subject to the weather patterns at that time and the artificial patterns of the Jordan River. Indeed, the level of the lake varies from year to year by over 1 m, with a negative final measurement, which produces a rapid retreat of the shores, especially the Israeli shoreline, which is less abrupt than the Jordanian shore. The tourist installations, constructed along the border of the sea in the 1960s, are today separated from the sea by several kilometers of muddy and salty beaches or salt deposits, which allow nothing to grow there. The lake, which once formed a single unit, is now divided into two parts. There is a northern basin, which is the larger and deeper of the two basins, where salt (halite) and gypsum occur as precipitates through concentration, and a southern basin that is shallower and covers at most a quarter of the total area, with saturated waters (up to 370 g/L), and functions as a large salt marsh from which salt, gypsum, potash and bromine are extracted.

The continuous decrease in the level of the Dead Sea, possibly compounded by the subsidence of the rift valley has serious consequences for the future of the lake (Nissenbaum 1993). Most notably, it has triggered a general resumption of the regressive erosion in the region downstream of small tributaries, resulting in the accelerated eating away of the banks into badlands and salt pillars that make the shores uneven and the water inaccessible. This is also accompanied by the deepening of groundwater tables and a decline in the springs. The loss of water over vast surfaces, water that cannot be used for drinking or irrigation, accentuates a process of desertification that is already quite advanced. The drying out of the Dead Sea is aggravated upstream by water being removed from the Jordan for various activities, thereby depriving the sea of its chief inflow, and downstream by water being pumped out (at the rate of 12 m^3/s) by extractive and chemical industries. The future of the Dead Sea looks rather bleak and will be accompanied by economic disasters related to tourism and industry. This realization has led to the proposal of a project jointly carried out by Jordan, Israel and the Palestinian National Authority, which would aim to raise the level of the Dead Sea by adding around 800 million m^3/year of water, which would come in from the Red Sea, through the Gulf of Aqaba (200 km). This project would have the advantage of halting the deepening of the groundwater tables and the harsh revival of the erosion process, would bring factories and tourist centers closer to the shores, and would bring the levels of the Aqaba and the Dead Sea almost 600 m closer, allowing for the installation of power stations, of which there is a very poor network in the region. It may even be possible to create a microclimate conducive to the hydration of vegetation. However, apart from the political conflicts and problems the region currently faces, there are other, significant stumbling blocks, and the cost of the operation, which is likely to be considerable, has not yet been clearly estimated. The water that would come in would be salty seawater, and would be added to the water in the Dead Sea. Consequently, there would probably need to be an extensive plan for desalination. However, the most notable difficulty is that the flow brought in would not, by itself, be able to maintain a volume of water that could support the development foreseen in terms of the demography, tourism and industrial exploitation of the site. Indeed, all things considered, the project remains a distant dream, rather than a possibility.

2.2.2.2. *Regions with hot desert aridity*

2.2.2.2.1. Lake Chad, Sahara (Sahel) 13°20′N, 14°0′E

Lake Chad, which is the fourth largest water body in Africa, with an area of 23,000 km^2 (in 1963), is the largest endorrheic lake in arid Africa (Bouchardeau and Lefèvre 1957; Schneider 1994). Unlike the lakes of Upper-Asia, contained in tectonically unstable intramountainous basins, Lake Chad occupies a large, open cratonic depression in Precambrian and Paleozoic rock.

The Chad basin is a flat and sandy region whose altitude varies between 300 and 500 m; it extends across over 2 million km^2, from the central Saharan region, in the north of the continent, as far as the humid, tropical savannas in the south. In the northwest, the basin is surrounded by ancient crystalline massifs (Precambrian) and volcanic massifs (Tertiary and Quaternary), the Aïr Ranges (2,000 m), the Hoggar (2,900 m) and the Tassili n'Ajjer sandstone-schist ridges (Paleozoic) (1,500–2,000 m). To the east, it is bordered by the high mountains of primary sandstones and recent volcanic ranges, namely the Tibesti (3,400 m), Ennedi (1,500 m) and Darfour (3,000 m). Raised banks at an altitude of between 300 and 500 m, which still retain the limestone streaks from Mid-Cretaceous marine transgressions, separate these ranges, through which the Chad basin communicates with the Libyan Murzuq and Kufra (<1,500 m). South, the border is more sinuous due to the multiple captures, or threat of capture, between neighboring rivers. This border weaves through the heights of the Ubangi-Shari and the Adamaoua, as far as the Benue Trough. The Benue is a tributary of the Niger and threatens to capture the flow of the Logone, thereby depriving the lake of this flow. Beyond this, through the Jos (1,700 m) and Zinder plateaus, the basin rejoins the Aïr. At the center of this formation are the large Ténéré and Fachi-Bilma ergs and Lake Chad, a type of *shallow lake* (Papon and Touchart 2003), with a large surface area.

This depression, which initially formed during the Oligocene (30 Ma), deepened and widened through subsidence during the rest of the Cenozoic and the start of the Quaternary (Pias 1970). It was gradually filled up, from south to north, by continental sediments from the destruction of surrounding rocks, with sometimes up to 600 m thick (Erhart 1953; Servant 1973; Gac 1980). The basic formations were as follows:

– at the base, Miocene sands from fluviatile alluvial fans with varying levels of clay. This is the equivalent of the Continental Terminal in the Sahara, covered by lacustrine deposits (hamadas) and fluvial and wind-blown sandy deposits of the Plio-Quaternary period;

– from the middle Pleistocene (around 700 ka), dunes formed by windblow sands,with intercalated lacustrine lens, and alluvial deltaic formations along the edges;

– finally, from the Holocene to the present day, from about 12,000 BP, lacustrine or deltaic deposits combined with the submerged and constantly reshaped eolian formations.

These formations are evidence of a climatic history marked by a major break in the ancient Quaternary (between 2 and 1 Ma), which was characterized by an increased intrusion of windblown sands compared to fluvial and lacustrine sediments (Garba *et al.* 2003). The end of the Quaternary period saw a largely arid ambience, made up of dry phases and wetter phases. The dry periods were those of the Last Glacial Maximum (20–15 kaBP) and several recurrences in the Tardiglacial (15–10 kaBP). They were marked by heavy deflation, which was made easier by the scattered green cover, and the presence of mobile sands that accumulated into windblown edifices (ergs). During the wetter periods, especially the Holocene Optimum (9.000–5,000 BP), sedimentation involved both coarse formations (>2 mm) brought in by rivers, and fine, sandy-clayey formations (<50 μm) brought in by the wind, and battered by the seasonal or occasional rains. The dunes, which were submerged, formed sandy banks or islands that were eroded and covered by vegetation. These structures are responsible for biochemical sedimentation (diatoms, emergent grasses) in lacustrine channels called *bahrs*.

Lake Chad appeared on the radar of researchers toward the end of the 19th Century, during colonial explorations. During this period, it was a large lake with an area of around 25,000 km^2, a depth of 5–8 m and an altitude of 284 m a.s.l. A few years later, in the early 20th Century, it had an altitude of only 280 m a.s.l. and had shrunk to a water body broken up by grassy banks, floating reed beds and archipelagos of sandy islands. As with all shallow lakes, Lake Chad has seen extreme variations in surface area and shape over time (Pouyaud and Colombani 1989). The maximum number of changes seem to have taken place around 9,000 BP, during the Holocene climatic optimum. This was the "Paleo-Chad" period, (Tilho 1928), a more or less mythical time when the lake is thought to have covered over 300,000 km^2, as far as the foot of the Tibesti mountains. Beaches (whose authenticity have, however, been contested) found at altitudes over 300 m a.s.l. even suggest that the lake may have communicated, beyond the raised banks seen today, with the large Libyan depressions, and could have emptied into the Atlantic through the Benue. From this one vast lake we only have a few residual lakes or ponds, sometimes brackish, and of course, Lake Chad itself. This lake has continued to shrink over the past millennium, alternating between high levels ("the Large Chad") at altitudes of 284, or even 286 m a.s.l., and low levels (the "Small Chad"), which can go as far as the near-disappearance of the lake. The high levels were observed in the 10th, 11th, 17th and mid-19th Centuries, while the low phases were observed in the late 15th and 16th Centuries, and the early 20th century (Maley 1993). The "normal", or most frequent states (the "Medium Chad"), are those around 282 m a.s.l. These alternations continue even today. The most recent were marked by a record extension in 1963 (altitude, 284 m; area; 23,000 km^2; depth, 4–6 m) and by near-total drying in 2001 (altitude, 275 m; area, 1,350 km^2).

The hydrology of the Chad (Olivry *et al.* 1996) is dominated by the climatic duality in the basin between the desertic north, ranging from arid to hyper-arid, and the tropical south, ranging from sub-arid to semi-arid. The north, below the Tropics, is the outer limit of the circulation of polar winter disturbances. Rainfall is rare, random, lower than 20 mm, and only very slightly more abundant in the mountains. The interannual gaps are enormous and there may be several consecutive years without water, with disproportionate evaporation of over 2,250 mm (Chevallier *et al.* 1985). There is meager vegetation, sparse or even completely absent. There is very limited, practically non-existent, chemical and biochemical alterations, and the fragmentation of rocks is a very slow process, producing clastites, pebbles, sand and silt. Rainfall is more abundant in the south, measuring over 400 or even 700 mm. The rains are also more regular between the dry seasons (exacerbated by the Harmattan that blows toward the SW in winter) and the wet seasons marked by the NE summer monsoons. The evaporation is slightly lower than in the north, but is still quite strong (about 2,200 mm). The vegetation in the south is denser and is characterized by savannas and gallery forests along rivers. Chemical alteration is the prevailing weathering process, with the destructuration of rocks producing silt and clay.

This climatic and geochemical duality can be seen in the transfer of matter from water input into the lake and within the lake itself. In older times, large wadis would come here from the Sahara. However, the progressive drying of the region over the last 7,000 years, at least, has deprived the wadis of water and they have been hidden below dunes and ergs, as in the case of the Ténéré region. Today, there is almost no water here apart from the floods in the Bahr el Ghazal, rising in the Ennedi and/or Tibesti Ranges, and this water is only available if it has not been completely removed for irrigation in Borku. The detritus carried by these wadis (clastites, moved through successive displacement through the floods), today, largely form inherited formations that are nearly immobilized and no longer come as far as the lake. The chief contribution of detrital material comes from the south, with 95% of it coming in on the Shari and its tributary, the Logona, which are tropical rivers with summer floods. These rivers carry a largely siliceous load, the majority of which are solid products of small caliber and abundant dissolved material. These conditions are reflected in the hydrological and sedimentological behavior of the lake, where it is possible to identify three distinct regions: in the south there is freshwater, or water with very low salinity, and biochemical siliceous (diatomaceous) sedimentation and clayey (largely kaolinite) sedimentation; in the north, we find brackish water that quickly dries out through evaporation, and detrital montmorillonite sedimentation; in the center is a complex, mixed zone with a neoformation of various smectites. Every year, the lake receives a total of around six billion m^3 of water through rain and 25–40 billionm^3 through rivers. However, it also loses about 46 billion m^3 of water through evaporation and infiltration. As a result, the hydrological balance is very fragile. (Pouyaud and Colombani 1989).

As it is so exposed, Lake Chad lost over 90% of its area and waters between 1963 and 2001. Around one-third of this was removed for irrigation and to sustain the thousands of Africans who depended on this water. The remainder was lost due to hydroclimatic reasons – the severe droughts toward the end of the 20th Century (especially the drought of 1984–1985) and the fact that the average flow rate of the Shari dropped to 213 m^3/s. Given these conditions, is it possible that there may be a total and persistent drying process which would cause the lake to disappear completely. This was a theory put forth by Tilho (1928), but it does not seem to have occurred in the historical period, as there have been an adequate number of years with rainfall to recharge the lake. Conversely, could we perhaps see the return of the vast Lake Chad from 8,000 or 9,000 years ago? This also seems unlikely, given that consecutive rain-deficit years are too frequent for this to occur. We must accept that the lake's geometry will vary around a "Medium Chad", with changing contours and physiognomy. However, it is known that there are two depressions, which are revealed when the water recedes. When the lake is full, like it was in 1963, it is a single water body, with two zones of free water along the eastern and western shores, and a median NW-SE strip of water with sandy ledges and islands. When the water level falls below 282 m a.s.l., as it did during the middle stage in 1997, two basins of free water appear to the north and south, bordered by sandy ledges and, along the eastern coast, by an archipelago made up of submerged dunes. Finally, when the lake is at its lowest, like it was in 2001, all that remains is a single basin of free water, in the south, at the mouth of the Shari delta. All of the remaining water, starting from the north, which is the first part to dry out, is dry or marshy, taken over by dunes, or prepared for agriculture. The lake's survival is, therefore, most dependent on the Chari–Logone flow. However, it is also determined by the hydraulic and agricultural installations that have been set up or planned in the region, and, more recently, by oil projects. If the lake does indeed disappear permanently one day, it will undoubtedly be due to human activities.

2.2.2.2.2. Lake Eyre, Australia 29°0′S, 137°15′E

Lake Eyre is the largest lake in Australia (Mabbutt 1977). It is comparable to Lake Chad in several aspects, however it also differs from Chad on notable points. Like Lake Chad, it is located in a very continental site, in a region with hot aridity; it is in the center of a flat basin surrounded by distant mountains. Like Lake Chad, Eyre is also the point of convergence and the lowest point in a drainage system that is partly tropical and partly desertic. Another similarity is that it is a shallow lake, or rather, a shallow sabkha, subject to large variations in volume, level and surface area. It varies between a large lake that is full (at these time, the Australians talk of an "inland sea") and a small, shrunken lake that is fragmented, dry and invaded to varying degrees by dunes. However, Lake Eyre is adapted to the bioclimatic and structural conditions that are unique to Eastern Australia, especially in its water supply or its interactions with groundwater and the hydraulic installations set up here.

The Lake Eyre basin covers around 1,140,000 km^2, which is close to three-fifths the size of the Chad basin, and one-seventh the size of the entire Australian continent. It lies in a topographical basin that is highly flared, a subsident depression that is a half-graben between a fault escarpment to the west (Twidale and Bourne 2004) and the Paleozoic Coastal Cordillera to the east. From the Triassic period up to the Quaternary period, this depression was filled with detrital sediments formed by continental quartz sand, which has become porous sandstone with intercalations of clay beds that have a partly marine origin. This is the aquifer sequence of Australia's Great Artesian Basin. The lake itself, strictly speaking, is located in the southwestern part of the depression, enclosed to the south by a tectonic threshold that blocks any outlet towards the sea. The lake's catchment basin is surrounded by relief features that are out of reach of even the highest waters. Far away to the northeast, there are ranges of flysch and limestone, injected by eruptive formations of the Great Dividing Range (1,000 to 1,300 m); to the north are sedimentary and volcanic hills, Mount Isa and Mount Barkly (<500 m). Finally, to the west is the crystalline Central Australian Plateau (500–1,000 m), broken up by E-W tectonic ranges: Mount McDonnell (1,347 m) and Mount Musgrave (1,440 m). The lake itself is located at the center of a *playa*, around 10 m below sea level, with many winding channels, sandy levees, marshy, clayey or salty stretches, yardangs and dunes.

E.J. Eyre was the first European to discover the lake in 1840, and it was explored and mapped extensively throughout the 19th and 20th Centuries; today, it is one of the best studied and tracked lakes in an arid zone (Kotwicki 1986). As with Lake Chad, it is governed by a dual climate that affects the behavior of its inflows. Its catchment area falls within a sort of Australian "sahel", which is subject to a large thermal and pluviometric gradient. The south enjoys a semi-arid, subtropical steppe climate, while the north enjoys a sub-humid tropical, climate, with over 500 mm of rain from the summer monsoon and an average temperature of 24°C. To the west and in the center is a region with tropical desert aridity, with random rain, resulting largely from convection and bringing in less than 150 mm, and an average temperature of 21°C. The lake is surrounded by desert. However, the Australian desert, which covers half of the continent, is not like the Sahara, although like the Sahara, it is one of the driest regions in the country, with evaporation rates of 2,500, or even 3,000 mm, in the center of the basin. This is a milder desert and with a slight increase in rainfall, it could transform into a semi-desert. The chief characteristic of the basin is the large difference in pluviometry and vegetation across both space and time (Hobbs, Lindesay, and Bridgman 1998).

Frankly arid zones, like the Simpson Desert, are flanked by steppes, where thousands of sheep and cattle graze. There are dry years, where the bare soil is exposed to continuous sunlight, and there are other years with heavy showers, where everything is covered and blooming, right up to the top of the dunes. While these conditions may have varied over time, they always did so within a climate that alternated between dry seasons or periods and wetter seasons or periods. This is a "rubefacient" climate, conducive to the fixation of the iron oxides that give the Australian plains their remarkable red tinge.

Lake Eyre (Kotwicki 1986) actually consists of two parts. To the north is a lake of around 8,400 km^2 with a shoreline that is about 9.5 m below sea level. The complex and changing contours of this lake enclose a flat bottom located at an average depth of 3.3 m. There is a deep hollow on its western coast, with a maximum depth of 5.7 m and a shoreline lying 15.2 m below sea level, which is the lowest point in Australia. To the south, another lake covers 1,200 km^2 with the shoreline lying at the same level, 9.5 m below sea level, with an average depth of 1.9 m and a maximum depth of 3.7 m on the southern coast. The Goyder canal, which is 15 km long with a width that varies from 200 to 1,500 m, forms a connecting channel between these two lakes. It allows water to be exchanged between them in either direction, depending on the direction of the wind and/or any difference in levels caused by floods. With the exception of the western shore, which has tectonic escarpments, the other shores are low and bordered by dunes or sandy cliffs shaped by the wind and through deflation. The flatness of the shores and the shallow depth of the lakes allow water to go back and forth easily over a wide floodplain, where a number of residual lakes persist after every flood.

As with all large lakes in an arid zone, Lake Eyre depends chiefly on water brought in from a network of tributaries. Like the networks that supply Lake Chad, the Lake Eyre network is asymmetrical. It includes wadis from the western desert and two large tributaries that rise in the tropical, humid mountains in the northeast: the Cooper Creek and the Diamantina River.

From the west, the wadis that indent the flexure of the large central plateau reach the lake with some difficulty, and are irregular. The largest inflow system, the Macumba River, drains into the northern lake, bringing in water that falls from Mount McDonnell and Mount Musgrave (600–800 km to the north). There is no inflow from the north: any flows that might arise are absorbed by the dunes in the Simpson Desert, where they feed the groundwater tables. In any case, the floods in all these wadis, which are quite small and ephemeral, die away quickly as they are diverted into multiple channels, and through evaporation and infiltration, making no noticeable contribution to the lake.

Unlike the Shari River, which flows directly into Lake Chad, the Cooper and Diamantina only reach Lake Eyre after a long journey through the deserts of Queensland and Southern Australia. While they are relatively large rivers in their upper basins, these rivers lose a lot of their force and regularity during this journey. In the east, the Cooper Creek has an average annual flow rate of 3.4 km^3/year as it emerges from a mountain basin that is around 150,000 km^2 large. It divides into multiple, interconnected channels on the glacis in the interior of Queensland. These channels combine with local flows and sometimes overflow in sheet floods. Two-thirds of the way into its course, the creek abruptly turns to the west, leading to the small Innamincka basin. Here, over a drainage area of 237,000 km^2, the average flow rate is no more than 2 km^3/year. However, in some years this may be as high as 12 km^3 with floodwaters over 3,500 m^3/s. Downstream, the Cooper Creek divides into two branches, which are unequally traveled by the floods. One branch moves to the south, toward Lake Frome and the other moves southwest, towards Lake Eyre. This is the branch that is less often traveled by floods. It is made up of an extensive network of braided channels, separated by very low interfluves that are unstable and constantly changing, with many dead arms, residual lakes and interdune ponds. Most of the time, the Cooper Creek does not go as far as the lake, as it is blocked by a barrage of dunes. On the other hand, it can overflow until it merges with the flow of the Diamantina, tens of kilometers to the north. At the center of the network of tributaries, the Diamantina opens directly into the northern part of Lake Eyre. At Birdsville, over a drainage area of around 115,000 km^2, after traveling over 700 km across the desert in a riverbed with multiple channels and an impressive width, the average annual flow is only 1.4 km^3/year; however, at its maximum, this flow can exceed 10 km^3/year, with surges of over 4,500 m^3/s. Thus, it is still characterized by the excessive, disproportionate nature of water courses in the desert:

The river! Surely the maddest, the most impossible river in the world, the Diamantina. It collects its mighty flood-waters from among the hills in north-central Queensland and rushes them down into the semi-desert of northeast South Australia. When this river runs, its width at Monkira is five miles. Here at Coolibah its width was fifteen miles, and at this time [end October] not a drop of water lay in the intertwining channels. The beams of the car's headlights swept up and then became level when the machine rose to cross each succeeding channel bank, and they reached down to light the channel beds for the car to take them, hugging the narrow track. The flood of the previous year had deposited grass and herbage seed, and now along many of the channels cattle and horse-feed grew profusely [...]. It was like a frozen sea, a sea frozen solid when in tempestuous anger; it was as though the car was a ship steaming over stationary waves. A. Upfield, *Wings Above the Diamantina*

Downstream of Birdsville, the floods spread out over the silty-sandy alluvial plains or are lost in a maze of playas, small lakes and marshy depressions. In this uncertain zone, the Diamantina receives the irregular contributions of the Eyre Creek, which collects and transports floodwaters from the large basin of the Georgina River (205,000 km^2; average annual flow 5.7 km^3/year; maximum flow 6.3 km^3/year and surges up to 3800 m^3/s), which flows along the edge of the Simpson Desert. The combined waters of the Georgina-Eyre-Diamantina then flow toward the northern lake through a network of channels without any slopes, with alternating currents that sometimes flow towards the lake, and sometimes away from the lake.

Lake Eyre thus survives through the rhythms of its inflows. However, these are not its only sources of water. While the lake cannot really depend on precipitation, with an annual average lower than 150 mm, which is not very significant, showers of moderate intensity may occur four or five times a year, and rainfall of 15–20 mm may produce a surge for an hour or two in the small wadis, while 50 mm can produce a flow for several days in a flood channel. We must also take into account the fact that the Lake Eyre basin overlaps to a great extent with the Australian Great Artesian Basin (Habermehl 1980), which discharges water into the lake in various ways. Through the swelling and upsurge of aquifers closest to the surface, or through springs that spout up, often related to tectonic accidents especially in the south or southwest, this water either flows directly into the lake or reaches the lake through short channels used by concentrated flows, or subsurface paths. Water also comes in through artificial discharge, from wells that have sunk several hundred meters into the artesian reserves. On the other hand, there are also considerable losses due to evaporation ($>$3,000 mm), infiltration and even the backflow of floodwaters in the tributary channels. All of this is difficult to measure and estimates often differ.

Geological evidence shows that toward the late-Pleistocene, between 45,000 and 25,000 BP, which was a cold and humid period preceding the Last Glacial Maximum, Lake Eyre extended over a large area. From this time onwards, until the present day, the climate has continually worsened: regional drying, shrinking water bodies and dominant winds. There have, of course, been temporary oscillations with humid episodes (for example, from 9,500 to 8,000 BP in the Holocene climatic optimum) and intermediate dry phases; there were also cyclical multi-year oscillations in water levels in the lake (floods and low water) and in its contours. The present climate of Lake Eyre may be compared to the climate in the driest periods in the last 10 millennia. Based on observations in the field (that go back to the middle of the 19th Century), the 19th Century was somewhat humid (and certainly wetter than the first half of the 20th Century) and the lake was often "full", i.e. a continuous water body. From 1945–1946 onwards, it became easier to track episodic changes using aerial photography and satellite images (Kotwicki 1986).

Toward the end of the 20th Century, the region saw some remarkable floods:

– The flood of 1949–1950 was the first to have been systematically tracked and analyzed. It started in April 1949, with the Cooper flowing over the dune barriers that blocked its mouth in the southern part of the northern lake. From November 1949 to September 1950, with inflows from the Diamantina and the small wadis around the lake, the volume of the lake increased from 2.5 to 17.5 km^3. In late September 1950, the lake was "full" and reached its highest shoreline of 10.3 m below sea level, with an area of 7,800 km^2. The floodwaters started receding in October 1950. The water retreated in the south first, and by 1953 the lake was almost completely dry.

– In 1974, the floods started in January, despite heavy losses through infiltration, as the water crossed the barrier made up of dunes. The northern part of the lake filled up first due to the Diamantina. From April onwards, the water entered the southern lake through the Goyder canal, along with (small) inflows from the Cooper Creek and the small, local wadis. In June 1974, the coastline was at 9.09 m below sea level. The floods started receding in August and continued to do so until June 1976.

– In December 1983, the lake was completely dry. There were heavy local rains from early January 1984 ($>$300 mm in seven days) that swelled the small tributaries to the west, which overflowed onto the floodplains and started filling in the depressions. By the end of January and early February, all the rivers and streams were in spate. The Cooper flooded the interdune hollows and managed to reach as far as the southern part of the northern lake (though with difficulty). This northern lake was eventually filled by both the surge from the Diamantina and the overflowing of the southern lake, through the Goyder Canal. By June 1984, both lakes were full and the floods had already started abating in the north. By February–March 1985, the entire system was drying out. Residual pools still occupied the gulfs along the southern shores of the northern lake. The southern lake was practically dry and salt deposits covered the gypseous clays.

These examples illustrate one of the most striking features of Lake Eyre's arid lacustrine system:

– the hydrological functioning of the lake is subject to the irregular inflows of water through rains and tributaries. The climate produces long, dry periods in which the only source of water are the artesian springs, especially in the south and southwest regions of the basin. The large floods in the Diamantina and the Cooper Creek, which result from tropical rains in the north and northeast, are severely reduced and are often even neutralized by infiltration and retention in the dune systems of the Simpson and Tirari Deserts. In addition to this, there is constant intense evaporation (>2,000 mm) across the entire basin;

– although the rains over the lake are not enough to trigger a flood, they are not negligible: 2.5 km^3 in 1949–1950, 8 km^3 in 1974–1976. However, the rainfall is still random;

– it is established. that the principal sources of water for the lake are the tributaries or inflows. In this respect, the Diamantina–Georgina–Eyre network plays a greater role than the Crooper Creek, both in terms of the volume brought in (38 km^3 against 6 in 1974) and since it has much easier access to the lake;

– depending on how rainfall is distributed over time and the arrivals of the floods, the lake can either be filled in from the north, like it was in 1974, or from the south, like it was in 1984. The exchange of water between the northern lake and the southern lake through the Goyder Canal can take place in either direction.

It has been estimated that a continuous, annual contribution of at least 20 km^3/year is required to ensure the lake has a constant presence of water. However, irregular floods, as well as the excessive pumping of water for human purposes, do not allow this today. Lake Eyre is an intermittent lake, which is only full twice or thrice in a century. When this does occur, the lake fills up relatively quickly (within a few weeks or months), depending on the intensity of the rains and floods. The flood starts abating almost immediately under the combined effect of infiltration and evaporation, but does so slowly, over a prolonged period. The lake starts drying from the north, which is hotter, flatter and has a higher rate of evaporation. The water stays longer in the south, especially in the gulfs along the southern shore of the northern lake. For reference, readers can consult the magnificent images of the 1974 flood in (Kotwicki 1986). It takes two to three years for the lake to undergo a complete "refill-drying out" cycle.

The long, dry periods exacerbate the concentration, through evaporation, of fine sediments in suspension and salts brought in by tributaries in spate and maritime winds. When empty, the lake is covered with salt deposits, 30–40 cm thick, over 90% of which is sodium chloride (NaCl). These salt crusts are swept away by deflation, which enlarges the surrounding dunes. Floods and receding waters on the flat bottom of the lake act on this salt (whose quantity is estimated to be hundreds of millions of tons), alternately dissolving it and concentrating it in an irregular rhythm, and the current transfers it from one end of the lake to the other. Given these conditions, it makes little sense to talk about the salt content in the lake; we can only speak of the quantity of salt stored in the lake, which could be around 500 million tons. Thus, in 1974 it was observed that 30 million tons of salt were displaced from the northern lake to the southern lake. In 1984, on the other hand, 40% of the salt was brought into the southern lake, 10 years before it was shifted back to the north through the Goyder Canal. For the majority of its existence, Lake Eyre has behaved more like a vast sabkha in a desert playa than a true lake.

2.2.2.3. *Regions with cold desert aridity*

2.2.2.3.1. Lake Balkhash, Central Asia 46°0′N, 74°0′E

Although it is located to the west of the mountains that separate China from Kazakhstan, Lake Balkhash has more in common with the lakes of Upper Asia than those in the Turanian depression (Létolle and Touchard 1998). Unlike the Aral or Caspian, this lake is not a remnant of the Sarmatian Sea, but is a true endorrheic lake perched at an altitude of 345 m a.s.l. Like its neighbors in Dzungaria and Xinjiang, the lake occupies the bottom of a tectonic depression, which is the westernmost and lowest of the large intra-mountainous depressions in Central Asia (George 1947). It was filled in from the Tertiary onwards by continental sediments, partly fluviatile and partly lacustrine (notably gypsum).

The catchment basin that supplies Lake Balkhash covers an area of about 500,000 km^2. To the north, it consists of a large plain that slopes downwards from a range of low mountains (1,500–1,600 m), and to the west there are low hills that separate it from the Aral basin. This part of the basin terminates in the lake, on a shoreline with cliffs. To the south and southeast, there is a large alluvial fan. The Ili River occupies the western part of this fan, made up of the deltas of several large rivers that arise in the high Alatau (4,000–5,000 m). The head of this cone makes up the southern shore of the lake, a low and marshy region. To the east, at an altitude of 400 m is the Dzungarian Gate, a mountain pass that connects the Turanian basin with the large topographic depressions in Upper Asia. The lake itself forms an oblong crescent 600 km long, divided into two by a strait that is less than 5 km wide and only a few meters deep. The western part, oriented SSW-NNE, is the wider part (up to 70 km) but is shallower (an average depth of 4.5 m). The eastern part, oriented W-E, is narrower (with a minimum width that is less than 10 km), but is deeper (an average depth of 7.5 m).

The lake is fed by precipitation, measuring around 100–200 mm/year, chiefly in the hot season (May–September) when the surface temperature may rise to around 30°C. These rains are offset, however, by an annual evaporation of about 1,000 mm. In winter (November–April), which can have very cold spells where the temperature goes as low as −40°C, the lake is covered by a layer of ice that is several decimeters thick, and evaporation is reduced. This pluvial-glacial-evaporative pattern is very irregular, varying through the year, as well as from one year to another. Indeed, the hydrological balance is only really maintained through the water brought in from tributaries. During the period between 1958 and 1961, considered to be a "wet" period, and pre-dating the construction of a large dam for irrigation on the Ili in 1969, the total water brought into the lake was 1,223 mm. Of this, precipitation accounted for 170 mm, the inflows accounted for 993 mm, and the groundwater discharge made up 60 mm. Evaporation in this same period removed 923 mm (Frecaut 1979). With a maximum flow in summer (over 900 m^3/s in July), the Ili River is the main inflow into Lake Balkhash. It supplies at least 80% of its water supply and most of its sedimentary alluvial load. Because of the Ili, the western basin has remarkably low salinity (1–3 g/L). The eastern basin, which is partly fed by the western basin through the strait but is otherwise lacking almost any other inflow, is subject to high evaporation and is much more brackish (3.5–4.5 g/L).

The most redoubtable factors, in a lake that is so flat and has an average depth of 5–6 m, are the significant changes in levels and area brought about by the variations in volume. Sedimentological and geomorphic evidence shows that in the past, there were more tributaries than today, that were also more powerful (Coque, Debaine, and Hazemann 2000).

The lake was much larger at times, especially during the Quaternary thaws, which were heavy toward the east, in the direction of the Dzungarian ports and the residual Sassyk-kul and Ala-kul Lakes. Some researchers even believed (although this is certainly wrong) that it could have overflowed into Upper Asia. At other times in history, on the other hand, the lake might have shrunk so much that it split into two separate lakes. Smaller changes, which were just as important, accompany the present-day hydro-climatic variations. Thus, between 1961 and 1987, which was a dry period (precipitation of 190 mm, evaporation of 1,010 mm), the level and the area decreased from 343 m and 19,000 km^2 to 341 m and 16,000 km^2. A more judicious regulation of irrigation, however, has allowed the level to rise slightly from 1990 onwards, despite a few small fluctuations.

2.2.2.3.2. Lob Nor, Upper Asia 40°20′N, 90°15′E

Central Upper Asia is characterized by the juxtaposition of vast topographic depressions, oriented W-E, surrounded by steep, mountainous ridges (George 1947). This is an example of what American geomorphologists call the *basins and ranges* system. The ranges (some of which measure more than 5,000 or even 7,000 m) have tiered vegetation, receive heavy precipitation, sometimes over 500 mm, and also support glaciers. They play the role of water towers, with respect to the basins, which are arid tectonic bowls (precipitation less than 100 mm) covered in steppes and deserts. The Tarim river basin, in the Chinese Xinjiang Province, is the largest of these basins. Closed off in the west by the Pamir Ranges, enclosed by the Tian Shan in the north and the Kunlun in the south, it contains the most hyper-arid continental desert in Asia, the Taklamakan, which is torrid in summer and freezing in winter, and whose dunes are very dense and entirely lacking in surface water. From the 12th Century onwards, Chinese maps placed a mythical lake at the eastern end of the desert, the Lob Nor, whose existence and discovery was the stuff of legend and controversy right until the dawn of the 20th Century (Létolle and Touchard 1998).

The first modern study into the existence of Lob Nor was carried out by Prjevalski (1879), following an expedition undertaken in 1876–1877. At this time, a lake (*the* lake?) was situated at the foot of the eastern end of the Kunlun Ranges, at an altitude of 780 m, in the southern part of the great terminal depression in the Tarim basin, which was taken over by mobile dunes. The Tarim, which skirts the Taklamakan erg on the north and east, used to bring freshwater into the lake from the Pamir and the Tian Shan. There was only a thin stream from the Kunlun Mountains. Swedish explorer, Sven Hedin, began exploring the north of the terminal depression in 1906 and, in 1911, revealed the presence of a lake situated 150 km northeast of the lake discovered by Prjevalski, and at an altitude of 750 m. This lake, which has no contact at all with the lake to the south, was only fed by a septentrional channel in the Tarim. The question that arose was whether these were two distinct lakes or one large lake that was split into two, or a "nomadic" lake that gradually shifted locations. From 1945 onwards, we have been able to get clearer answers to these questions because of aerial surveillance and satellite images (Zhao and Xia 1984).

It is clear that the flat lake, with an average depth of barely 5 m, has undergone large variations in area over time. One of the notable pieces of evidence for this is the existence of several regression terraces downstream. Memories of an "inland sea" (some claim it covered an area of 20,000 km^2) have persisted in China's collective memory since Antiquity. Toward 6,000 BCE, the terminal bowl in the Tarim basin was covered by a water body measuring 4,500 km^2. The lake described by Sven Hedin measured 80 × 40 km. In 1991, it measured only 29 × 15 km, a little over 3,000 km^2, with a depth of 5 m.

The Lob Nor is a good example of an unstable lake in an arid region around the middle latitudes. The contours, surface and even the position of such lakes vary according to the vicissitudes of the climate and the waters brought in by their inflows. The Lob Nor suffered the direct effects of the tectonic and climatic changes that accompanied the post-Wurm thaw and the progressive aridification of the region: subsidence, silting, wind morphogenesis and fluviatile meandering. Today the basin receives precipitation in the form of summer rains (May–August), which are irregular and less than 50 mm, compared to at least 2 m of evaporation, even when reduced by the water freezing in the winter (November–March). The lake would not have survived without the input provided by the torrents that descend from the mountains and are collected over more than 1,200 km by the Tarim River.

At present, from the start of the 20th Century onwards, the lake observed by Prjevalski is only reached by exceptional floods in the Tarim. Ordinarily, floods take a northern branch (with a SE orientation) of the river towards Lake Hedin (750 m). Depending on how the flow varies, the lake (or lakes) either extends across large floodplains, or recedes, leaving residual lakes and vast surfaces open to deflation and the cutting away of the yardangs and formation of the dunes, all of which form obstacles, obstructions or diversions for incoming waters.

The Lob Nor has relatively low salinity. Prjevalski said that the lake he discovered had been a freshwater lake. A salinity value of 5 g/L was estimated. However, it is likely to have varied considerably based on the contribution from the Tarim, ranging from brackish salinity to the salinity of a hyper-arid sabkha.

Today the lake is poorly and irregularly fed, dried out by evaporation and an increasing diversion of the waters of the Tarim for irrigation, to feed the growing population, for the oil industry, and so on. Based on satellite images, the Lob Nor appears to be a dying lake, moribund and nearly dried out.

2.2.2.4. *Arid regions in the mountains*

2.2.2.4.1. Issyk-Kul, Upper Asia 42°25′N, 77°15′E

Lake Issyk-Kul is a mountain lake in Upper Asia (Savvaitova and Petr 1992; Létolle and Touchard 1998), whose characteristics are closer to those of Lake Baikal than to Lake Balkhash. It is located at an altitude of 1,606 m in a tectonic rift oriented W-E, surrounded by the glacier-covered peaks between the Tian Shan (5,200 m) to the south and the Alatau (4,770 m) to the north. It is one of the largest lakes in Asia, the sixth largest by area, covering 6,236 km^2, and third largest by volume, with 1,738 km^3. It has an average depth of 279 m and a maximum depth of 668 m, partly attributable to the deepening in the Quaternary glacial.

Through its thermal inertia, this mass of water can create an ambient microclimate that turns the borders of the lake into a sort of oasis in a region with cold, sub-desert aridity at a high altitude. In winter, the temperatures are low (–16°C in January), but may be 4–10°C higher than the regional averages. Summer temperatures are attenuated (17°C in July) by altitude and the cool lake waters. There is an average rainfall of 250 mm, distributed between rainfall through convection in summer, and cyclonic rain in winter, with pronounced dry spells in autumn and spring. Evaporation, which removes about 700 mm of water, is active in all seasons through the eastern or western winds that are channeled by the narrow corridor that contains the lake.

In addition to localized rains, Lake Issyk-Kul is fed with up to 15–18 million m^3/year by a catchment basin with an area of 15,800 km^2, which receives several small streams from the neighboring glaciers, and from groundwater discharge. There is an outlet in the west that is only active when the lake level touches an altitude of 1,620 m. The water that overflows then follows a very flat canal that is 6 km long and empties into the Chu river, a former tributary of the Syr Darya, which is today lost in the sands of the Aral basin. This was the situation at the start of the 19th Century, when the precipitation increased to at least 300 mm, evaporation was 630 mm and the lake also received the equivalent of 600 mm through its affluents. These conditions disappeared in 1820. For the rest of the 19th Century and in the early-20th Century, the spillway canal continued to function (Berg 1932), but flowed in the opposite direction acting as an outlet for the floods in the Chu, leading towards the lake. It has been completely abandoned since 1930. From then onwards, given the decreasing precipitation and the retreating glaciers, the floods steadily diminished in volume and the lake level continued to fall. It was 1,609 m in 1930, 1,608 m in 1950 and 1,606 m in 1985. Today, the lake is completely isolated.

This regression has drastically changed the lake's hydrological behavior. The floods that form in the hot season (from the end of spring to the end of summer) always bring about an interannual variation in the level of the lake, in the order of 25–30 cm, in an environment of general dryness. However, the overall water balance remains negative and is further worsened by the development of irrigation. With the outlet dying away, the drainage that ensured salts were removed no longer existed and the lake became brackish, with a sodium sulfate salinity of 5.8 g/L, distributed quite uniformly across the whole water body. The sedimentary load of the inflows, estimated to be 900,000 t/year, is also no longer removed and the solid and dissolved material brought in accumulates entirely within the lake. Like all large lakes, the Issyk-Kul has its own currents, which, when reinforced by the wind, lead to a superficial mixing of the waters. However, the principal factor governing how water is distributed through the depth is temperature. In summer, the average temperature on the surface and up to a depth of 15 m, is around 20°C, while the temperature at the bottom is around 4°C. There is, therefore, the normal stratification of warm water on cold water and no vertical mixing. In winter, it is chiefly the shores and the gulfs forming digitations inside the land that are frozen. The temperature on the surface of the lake rarely goes below 3.8°C, which is close to the temperature at the deepest point of the lake. The resulting homothermy extends across the layers, allowing an effective mixing of waters and oxygenation, which sustains life even in the deepest parts.

The major problem here is to stop the worrying drop in the level of the lake, especially as Lake Issyk-Kul, like many lakes in an arid zone, attracts an increasing population that lives on its shores, new activities and an increasing demand for a water supply. There is already over 30 m^3/s of water pumped out of the lake and its inflows in the summer for irrigation purposes. And while polluting discharges may not be too severe yet, there is no doubt that in just a few years they will be a source of worry.

2.2.2.4.2. The Great Salt Lake, Utah, North America 41°10′N, 112°30′O

The Great Salt Lake in Utah is the only lake in the arid regions of North America that bears comparison (in terms of dimensions, history and hydrology) to the large, intramountainous lakes of Asia as well as the large sabkhas in Africa and Australia. Like these other water bodies, it is a remnant of quaternary paleo-lakes that were once much larger than the present-day lake. And like many of the large lakes on other continents, it is a flat, saltwater lake, a shallow lake or large sabkha, lying in a playa. Trappers came across the lake towards the end of the 18th Century and the Great Salt Lake has been the subject of several scientific studies from the 1850's onwards (Stansbury 1852).

It is located in the northeastern part of the Great Basin, a series of high plains lying between the Rocky Mountains and the coastal Sierras, broken up into multiple sections (*basins*, at an altitude of between 1200 and 2,000 m), separated by ranges that are over 3,000 m high, and cut by glaciers. In the Pleistocene, most of these basins were occupied by lakes, some of which were very large, such as Lake Lahontan (22,000 km^2) in the northwest and Lake Bonneville (52,000 km^2) in the northeast (Currey 1990).

This latter body is the ancestor of the present-day Great Salt Lake. Evidence of its existence throughout the Pleistocene period is sedimentological (fossil deltas, lacustrine deposits) as well as geomorphological (perched terraces) in nature (Gilbert 1890). The oldest traces of this ancestor lake, dating between 700,000 and 300,000 BP, are found at around 1570 m a.s.l., i.e. nearly 300 m above the present location of the lake. On several occasions, in this period, the glacier barriers that held back the lake were broken and the water was able to flow as far as the ocean through the network of the Snake and Columbia Rivers. Another maximum height has been identified, attained during the recent Pleistocene, around 15,000 BP, at 1,560 m and over an area of 51,800 km^2 (Abrahams and Parsons 1994).

This maximum height was followed by a series of hot and dry episodes, with lakes shrinking and strong wind deflation, alternating with cold and humid episodes with the lakes expanding, and accompanied by erosion and sedimentation. After the episodes, the lakes were located at increasingly lower altitudes. Around 4,000 BP, the lake was still situated at 1,298 m a.s.l., which was 20 m higher than its present-day location. In 1873, it was only at 1,283 m, covering an area of 6,200 km^2. From this time onwards, the level has decreased again. It reached its historical minimum in 1963, at 1,278 m, over an area of 2,400 km^2, with an average depth of 8 m.

The vast intramountainous depression in which the Great Salt Lake is located is cut off in the east by the overlapping mountains of the Wasatch Range (over 3,000 m high), a western front of the Rocky Mountains (Aubouin *et al.* 1986).

This depression itself is broken up by parallel ridges, oriented N-S, and shaped by glacial and desert erosion until they are sometimes reduced to inselbergs. One of these ridges cuts across the middle of the lake and divides it into two basins. The northwest basin is larger and deeper than the southeast basin. The Great Salt Lake Desert extends to the west of the lake, an arid plain at an altitude of 1,200–1,300 m, which was the bottom of the ancient Lake Bonneville, transformed today into a vast, salt-encrusted playa. The lake itself, at its maximum dimensions, is a water body that extends over 116 km from the northwest to the southeast and is around 35 km wide.

The Great Salt Lake, as it is at present (Houghton 1994), no longer has an outlet like Lake Bonneville did, however it is fed by the same kind of inflows. The major part of the water brought in (around three-fifth or close to 2 km^3/year) comes through a network of tributaries issuing from the Uinta Mountains, whose highest point is Kings Peak (4,114 m). To the north, the winding Bear River, a former tributary of Lake Bonneville when it was 20 m higher, has its headwaters 145 km from its mouth into the lake, but it takes a detour of 800 km in order to go around the Wasatch Mountains, before finally reaching the lake. En route, the river receives water from Lake Bear, a glacier-dammed lake that is artificially maintained at a constant level. In the center, the Weber River goes directly to the lake, while the Provo Lake first feeds Lake Utah, which in turn feeds into the Great Salt Lake through the Jordan River. All these water courses are nival courses, with spring floods (May–June) and low waters in the winter. Other inflows of water (around 1,25 km^3/year) come in from groundwater discharge and springs, including karst springs, but chiefly from precipitation. The rainfall that occurs directly on the lake is rather low, often less than 200 mm, but it can go up to 300 or 400 mm on the western face of the neighboring mountain ranges and fall in the form of snow. Precipitation is concentrated into a highly restricted number of days (less than 50 days) and occurs mid-season, in the form of storms or heavy showers. On the other hand, there are constant losses through infiltration, which are difficult to measure, while losses through evaporation are not much stronger in summer than in winter, despite the large thermal difference (25°C in July, -1.9 ° in January). The total seasonal variations in level are minimal, being around 30 cm.

The Great Salt Lake, a shallow lake, receiving only a mediocre supply is, as the name indicates, highly saline. This is its legacy from a long endorrheic past. In the

1870s, the earliest measurements carried out to establish salinity found this to be 150 g/L and found that the chemical composition of the lake is close to that of the ocean. Today, the salinity of the lake is 270 g/L (128 g/L in the southern lake and 332 g/L in the northern lake), which prevents the lake from freezing over despite the low winter temperatures. The salinity does, however, vary according to the season, location and depth. A change in level of even a few decimeters can cause the salinity to vary between one and three percent in either direction.

It is estimated that the lake contains a permanent mass of about 5 billion tons of evaporites, largely NaCl, maintained on an annual basis by 3 – 5 million tons of minerals in solution. The lake also receives a large volume of solid sediments, mainly clays and silt. When dry, along the exposed foreshores and the dried out bottom of the playa, criss-crossed by polygonal desiccation cracks, the surfaces are so hard that they can easily support heavy vehicles. They are also flat, which makes them an easy choice for competitions and attempts to set records for automobile speed.

In the context of a mountainous region in the middle latitudes, Utah's Great Salt Lake is a lake in cold, sub-desert aridity, like the lakes of Upper Asia.

2.2.2.4.3. Lake Titicaca, South America 15°50′S, 69°20′O

Lake Titicaca (Dejoux and Iltis 1992; Dejoux 1994) has the unique distinction of being both a high-altitude lake as well as a tropical lake. Located at the latitude 15°50′S, it is almost symmetrical with Lake Chad (13°20′N) with respect to the equator. However, it is found at an altitude of 3,809 m a.s.l. and in a sub-desert environment that verges on the sub-arid (Meigs 1977–1979). It is the largest lake in the Andes: 180 km long, running northwest to southeast, and 50–60 km wide, with an area of around 8,500 km^2 (in 1994).

Geologically, Lake Titicaca's position is comparable to that of the Great Salt Lake in Utah. Like the latter, it is situated at the heart of a recent mountain chain (Late Cretaceous–early Tertiary), in the northeastern part of a vast intra-mountainous depression, at an average altitude of close to 4,000 m, covered by Plio-Pleistocene sediments. This depression is the Altiplano, a high plain broken up by ridges and basins, lying in a N-S orientation between two cordilleras. One is a coastal range to the west, around 5,000–6,000 m high, composed of marl-limestone and volcanic in nature; the other is an inland range to the east, whose highest point is 6,500 m, composed of marine marl-limestone as well as continental formations such as sandstones, schists, gypsum and volcanic rocks. This is the equivalent of the Great Basin in North America, but is thrice as narrow and thrice as high. At the start of the Quaternary, after the initial glacials (1.5 Ma), almost the entire Altiplano was covered by a large lake, Lake Mataro, as evidenced by sedimentary deposits at an altitude of 3,950 m. During the Pleistocene, every glaciation (there were four in total) was followed by the formation of lakes, which grew smaller and smaller in size and appeared at increasingly lower altitudes. The last of these lakes was Lake Tauca, located at 3,820 m, which dates to approximately 13,000 BP and is considered the most recent of the Titicaca Paleolakes.

The present-day lake is the chief feature of the major Peruvian-Bolivian Altiplano endorrheic system. This includes Lake Titicaca and a lake formed through its outlet, 280 km further south, Lake Poopó. Lake Titicaca consists of two unequal basins, separated by a narrow strait 850 m wide and 20 m deep, which goes through a transversal ridge. The northern basin has an area of around 6,550 km^2 and a maximum depth of 280 m, while the southern basin measures at least 1,450 km^2 and has a maximum depth of 40 m. The eastern shore of the lake, which is contiguous with the eastern cordillera, is abrupt and is broken up by faults. The western shore, which is bordered by the Altiplano, is low and indented by shallow bays that are more or less closed and crowded with half-submerged reed beds.

Because of the latitude which it lies on, Titicaca falls in a climatic region where seasonal variations in temperature are smaller than diurnal variations. The annual difference in temperature is lower than 6°C, however due to the altitude, the annual average is quite low, around 8–10°C. On the other hand, because of the sunlight it receives, daytime temperatures can easily exceed 20°C, while frost frequently occurs at night or dawn. However, due to the thermal inertia of the mass of the lake, the average temperature of the water is maintained at around 13°C. Consequently, even when frost sets in at night, ice does not form on the water except in shallow bays, and even then, only for a few hours. It is hard to measure the precipitation occurring directly over the lake given the lack of an adequate observation network. Some of the precipitation is brought in by westerly winds from the Pacific that travel through the coastal cordillera, which are high and hard to cross. This precipitation results in between 500 and 600 mm of water falling on the western face and the peak of the mountains. However, it brings in less than 250 mm over the Altiplano. The major source of atmospheric humidity over the Altiplano instead comes from the Amazon forest through convection over the eastern cordillera (Hobbs, Lindesay, and Bridgman 1998).

The total annual rainfall around the lake is between 250 and 500 mm, and occurs primarily toward the end of spring (November) and early autumn (February). In keeping with tropical patterns, rainfall also occurs in the form of afternoon or nocturnal storms. On the other hand, under a generally clear sky, evaporation can be as high as 1,500 mm per year. Losses through lateral and sublacustrine infiltration are also quite large due to the large number of faults and cracks present.

The lake's hydrological balance depends more on precipitation than on tributaries. This is because there are very few rivers, and they arise largely in the eastern cordillera, with an annual flow that is little more than $270\,m^3/s$ and which may decrease further with the progressive disappearance of present-day glaciers. The inflows are balanced by an outflow in the southern part of the smaller basin, the Rio Desaguadero. When the water level of the lake is high, this canal carries away the excess water to Lake Poopó and the neighboring *salinas*. When water in the lake is low and when it rains over the Altiplano, this canal may carry water in the opposite direction, towards the lake. Indeed, while the annual variations in levels are relatively small, being less than one meter per year, interannual oscillations can be larger. As a result of these variations, the western and southwest shores, which are the flattest of the lake's shores, are exposed to either large floods (as occurred in 1986) or to droughts that threaten the survival of aquatic vegetation (as occurred in 1982–1983). The relative stability of Titicaca's hydrological pattern is also reflected in its hydrochemical balance. The lake's pH varies over a very narrow range, 8–9 at most, and the mineralization is around 1 g/L. Given the large presence of limestone and alkaline volcanic rocks, the hydrochemical facies is sodium-chloride and sodium-sulfate. However, the high pH favors a large concentration of silica, especially in the deeper waters where diatom tests are abundant. The predominance of the diurnal variation of temperatures (a feature of tropical climates) and the relatively low temperatures because of the altitude, result in the cooler, heavier layers of the water near the surface sinking down, thus contributing to a continuous vertical mixing of water and the oxygenation of water throughout the profile, especially in winter. These circumstances promote the abundant formation of algae and macrophytes (Characeae and reeds) and a rich variety of ichtyological fauna.

With the exception of the large inland "Seas", the Caspian and Aral, no other lake in an arid region has such a dense population on its shores [ref. the *World Population Distribution Map* at 1/15,000,000, de G. DECROIX, (In Noin 1997)] or makes a greater contribution to local economic life. However, like all lakes in the Altiplano, the future of Lake Titicaca seems uncertain. Several of these lakes have already ceased being lakes, becoming sabkhas (*salinas*) or salt deserts (*salares*), such as the Uyuni Salar in the southern part of the Altiplano, a vast, dry lake at an altitude of 3,800 m, which is now a salt reserve and whose salt blocks are even used for the construction of houses.

2.2.2.4.4. Lakes in the Tibetan Plateau, Upper Asia 30°0′ to 38°0′N, 80°0′ to 100°0′E

The Tibetan Plateau, "the highest desert on Earth", is located at an average altitude of 4,800 m a.s.l. between snow-covered mountains that are higher than 7,000 or even 8,000 m: the Himalayas to the south, the Kunlun to the north and the Karakoram Ranges to the west (Fort and Dollfus 1992). The plateau has an area of approximately 2 million km^2, about four times the size of France. It is closed off by massif chains oriented W-E, with rounded peaks, despite being at 5,000 or 6,000 m a.s.l., which dominate poorly drained valleys and high basins dotted with lakes (Létolle and Touchard 1998). These are mountainous blocks within a subduction system that push the Indian continental plate below the Asian plate. This movement, which began in the early Eocene, 50 million years ago, continues even today and manifests itself through multiple earthquakes, a slow but continuous elevation of the plateau, by the slow displacement of Tibet to the northeast at the rate of 2–3 cm/yr, and through active tectonics producing breaks and lateral faults (slip-strike faults) (Mascle, Delcaillau, and Hérail 1990; Debelmas, Mascle, and Basile 2008). Topographically depressed with respect to the high mountains that surround it, Tibet is overall endorrheic, except in the southwest where it is open to the high Indus catchment, and in the east, where water flows toward China and the Bay of Bengal. This position in a depression, but at a high altitude, is in some respects similar to some features of the Andean Altiplano (Francou 1993) and the North American Great Basin, and explains to a great extent the aridity and the lacustrine nature of the Tibetan Plateau.

Tibet is a dry land – one of the driest mountainous regions in Asia – due to its sheltered position. In the south, the Himalayas form an obstacle that prevents the Indian monsoon from coming in; to the east, the Chinese monsoon does not enter because of how deeply continental the region is; in the north and the west, residual flows of polar air through the Kunlun and Karakoram prevail. In addition to all of this is the fact that at these altitudes the air holds very little water vapor, and the presence of a strong thermal anticyclone in winter significantly decreases any chance of precipitation. Rainfall is low and convection or storms occur chiefly in the summer; snow rarely occurs, when it does, it is at the end of summer or early autumn. Despite the paucity of tracking data, we know that snowfall across the plateau is less than 200 mm and barely more than 400 mm in the surrounding mountains. Because of its high altitude, and despite being toward the middle latitudes, temperatures are low at the 4,000 m mark: an annual average of between −1 and −6°C, and average temperatures of −12°C in January and 10°C in July. However, in the extremely dry air, evaporation (or sublimation) remains very large: around 1,500–2,000 mm/year. This is, thus, a region with cold aridity and temperate summers.

Given these conditions, how can the presence of so many lakes on this desert plateau be explained, if not by the large number of topographical basins that act as receptacles for the water in the mountains? These basins are the chief feature of the region's morphology (Fort and Dollfus 1992). Many of them are of tectonic origin: syncline folds, grabens or endorrheic depressions between parallel faults that are still active. Other basins are more strongly linked to glacial, periglacial and wind dynamics, all of which are prevailing factors across Tibet. Ancient glaciers have left behind many traces: natural amphitheaters high on the slopes, U-shaped glaciated valleys, overdeepened basins, abandoned moraines and fluvioglacial sedimentary deposits. Wind dynamics have always played a key role on this high, bare and dry plateau: a layer of dispersed quaternary loess, deflation basins, eolian constructions. This unique evolution, which saw the combined effects of tectonic mobility and climatic changes, led to large disruptions in, and disorganized the hydrographic network: water courses being diverted or interrupted, the capture of rivers, the creation of endorrheic spaces through tectonic blockages or obstacles created by moraines or dunes.

Tibet, and Western Tibet in particular, became a veritable field of lakes, most of which are still only poorly known (Létolle and Touchard 1998).

We can still try, however, to pick out a few traits that are common to all these lakes. All the present-day lakes are remnants of a lacustrine past when lakes were spread out over a much larger area, perhaps four times as large as the area covered today. Following each glacial (there were three or four during the Quaternary according to Chinese authors), the lakes reached higher levels. For example, after the Last Maximum Glacial (around 14,000 BP), and at the start of the Holocene (around 9,000 BP). Evidence of their presence has been found at altitudes over 5,000 m in the form of shorelines, beaches or simply significant sediments. It is certain that there were fewer lakes, but they were larger and deeper, fed by melting snow. During the final glacial conditions (Younger Dryas, 10,000 BP), followed by the progressive drying of the climate, the lakes broke up and shrank. Low levels mark the boundary between the Pleistocene and the Holocene. The present-day lakes are greater in number, widely scattered, shallow and more or less blocked up by fluvio-periglacial sediments.

The hydrology of Tibetan Lakes reveals the effects of relief and altitude on water regime and the temperature of water in a cold continental desert. These lakes receive water largely through rain and melting ice. Precipitation is usually low and in the case of excess precipitation, is unevenly distributed. Certain basins receive less than 50 mm or even less than 25 mm of precipitation. The southern slopes receive the most water through heavy and sudden showers. Debates are still ongoing over whether the ancient glaciations were dispersed over several sites, as several Chinese authors believe, or whether there was a single ice-sheet that covered the entire plateau. Regardless of which theory is right, present-day glaciers are fed by summer snows and are all thin and perched, almost immobile, resting on permafrost. At the beginning of summer, these glaciers start weeping, and the surface of the permafrost across the slopes, which forms a more or less continuous cover, undergoes thawing. This results in surface flows, which are maintained by some springs and which may be diffused or concentrated into a number of streams with low flow. Lake Namtso, northwest of Lhasa, has 156 water bodies that flow into it, bringing in only 1 or 1.5 km^3/year in total (Létolle and Touchard 1998).

Annual variations in levels are rarely greater than 10 cm. Inflows carry coarse sediments, sands and pebbles into the lakes and these sediments slowly choke the lakes or are lost in the marshy depressions where the vegetation is concentrated.

Lakes in Tibet freeze over in winter for four to five months, from November to March or between December and April, upto a depth of 30–40 cm on the surface. In summer, thermal stratification returns to normal, with warm water (15 or 16°C) lying on the cold water (3–9°C at the bottom). In autumn and in spring, when the surface waters cool or grow warmer, respectively, there are two periods of mixing in the depths, with homothermy through the vertical profile. Salinity is usually low, despite an evaporation-sublimation rate that is estimated to be 2000 mm/year, although this value could be an overestimation. Measurements are quite scarce, but the salt content (Ca and especially Mg, halite and gypsum) seems to increase with aridity from the south (1,7 g/L with a pH of 9.4 in Lake Namtso, located at 31°N and an altitude of 4,627 m) to the north (12–15 g/L with a pH of 6–8 in Lake Koko Nor, or Lake Tso Ngömpo, situated at 37°N and an altitude of 3200 m, at the eastern end of the Tsaidam basin).

The lakes on the Tibetan Plateau are good representatives of high mountain lakes in the cold and arid band along the middle latitudes in Asia.

2.3. The principal features of lakes in an arid zone

The examples that we have seen allow us to identify some basic features of lacustrine systems in deserts.

All the lakes belong to hydrographic networks that are endorrheic, topographically closed, without any outlet for water except into the atmosphere (through evaporation) or into the ground (infiltration). Many of them are mountain lakes, located in hollows in rugged relief. However, others are lakes located in plains or plateaus, formed on ancient surfaces with mild deformations. The majority of the lakes lie in depressions of tectonic origin: subsident depressions (Lake Chad), intra-mountainous syncline basins (Balkhash, Lob Nor), or rifts (the Dead Sea, Issyk-Kul). However, there are also some lakes that occupy alveoles produced by or deepened by ancient or present-day morphodynamic systems: drainage through hydro-eolian activity, glacial over-deepening (Issyk-Kul), fluviatile dams (Sea of Galilee, Balkhash), moraines (Tibet) or dunal systems (Lob Nor) where water has accumulated on the sheltered side. Some lakes belong to endorrheic systems made up of multiple basins aligned in a chain (Tiberias - Dead Sea), connected in a network (Great Salt Lake–Lake Bear–Lake Utah; Lake Titicaca–Lake Poopó), or scattered over a region (the Tibetan Lakes). The position of lakes in an arid zone, like those in other regions, depends chiefly on the structural context, especially for sites created by Meso-Cenozoic orogenesis and its Quaternary sequels (neotectonics), as well as local geomorphological evolution. However, the characteristics of each lake are determined by past climatic variations and present-day aridity.

Almost all lakes in arid zones are situated in regions with sub-desert or desert aridity, whether hot or cold aridity, and whether on plains or mountains. These regions are often at the outer limit of sub-aridity. Lakes in hyper-arid regions are rare or one-off exceptions. The largest of these had much larger expanses in the past than they do today. They were preceded by enormous paleolakes (Lisan, Bonneville, Mataro), that at times appear mythical (Paleo-Chad, Lob Nor). The maximal expanse of these paleolakes occurred at several instances during the Quaternary, undoubtedly after each glaciation, and especially during the Last Great Interglacial (around 125 kaBP) and during the Holocene climatic optimum (around 9 or 8 kaBP) (Petit-Maire 2002).

However, from this time onwards there was a widespread shrinking of lakes. Many lakes disappeared altogether (the Saharan Paleolakes, Taoudenni, Timbuktu, the South African Etosha-pan and Okavango-pan), or are on the verge of disappearing (Lob Nor), which corroborates the common planetary history of the climate drying out in a progressive, if irregular, manner. In any case, this shrinking of lakes had several significant effects. Some lakes lost their exorrheic outlet (Balkhash, the Great Salt Lake). Dead lakes are now only represented through fossil deposits (lacustrine or deltaic), through shorelines or perched terraces.

The contraction of the lakes led to a reduction in volume, a drop in the altitude of the water body and, consequently, reduced the depth based on the topography of the bottom. The deepest lakes, located in rifts or unstable depressions, with a depth of over several hundred meters (the Dead Sea, Issyk-Kul, Caspian Sea) generally endured as a single water body. Others, which used to be a single contiguous body, were divided into several basins separated by banks and connected through straits (Aral Sea, Balkhash, Great Salt Lake, Eyre, Titicaca), or were even divided into isolated lakes (Lob Nor). Those that occupied basins with flared bottoms became flat lakes, with a depth of less than 50 m or, sometimes, even less than 10 m (Aral Sea, Balkhash, Chad, Eyre), revealing silty-clayey surfaces, with varying levels of salt, along their margins. Vegetation (and sometimes cultivated crops) have tried to colonize these surfaces. This tendency to shrink is still in effect, although, after the great droughts at the end of the 20th Century, a certain stability has been observed, with a slight resurgence in some cases. Human intervention, which is becoming more frequent, plays a role that is both difficult and ambiguous. On one hand, population growth, overgrazing and over-exploitation of agricultural land, the increase in irrigation, industry and in tourism all contribute to the overexploitation of lacustrine reserves (Aral Sea, Dead Sea). On the other hand, a rational arrangement of dams and protective installations, national regulation and international agreements all attempt to preserve, if not replenish, the availability of water, an indispensable resource for economic and human development.

Because of their endorrheic nature, all arid lakes, with their catchment basins, make up a closed and independent hydroclimatic system (Frecaut 1979) characterized by its water supply, its losses, its chemical and sedimentological compositions, biological capacities and by how vital cycles occur across time scales.

From the point of view of how these lakes are fed, the direct supply through rains is primordial, especially for smallest lakes that have no inflows. However, the rainfall is always poor (less than 200 mm) and very irregular. But while they might not be capable of triggering a flood, the direct rains are far from being negligible and may supply the lake with between three and five percent, or even up to 20% of the total water it receives. In addition, if the catchment basin contains relief features that are sufficiently elevated and receive good (or better) levels of precipitations, these can act as water towers (Issyk-Kul, the Great Salt Lake, Titicaca); or if the catchment basin is large enough to receive distant precipitation that is more abundant (Chad, Eyre), the lake can then benefit from up to 80 or 85% of additional inflows that may compensate for local insufficiency. In these cases, the tributaries are the principal sources of water. However, these inflows are always irregular and uncertain, given the existence of long dry periods, which exceed six to seven months everywhere, and can even last longer than 12 months in the most extreme cases (Lob Nor, Eyre). On the other hand, losses through evaporation are significant everywhere, almost always being more than 1,000 or even 2,000 mm (Chad, Eyre, Lob Nor), including arid lakes in temperate regions with hot summers and cold winters (Caspian Sea, Aral Sea, Balkhash). The relations of these lakes with underground circulation are harder to define and measure. They clearly depend on the nature, structure and topography of the terrain. Groundwater discharge through gravity or from karst springs is generally larger on slopes than on plains, but artesian upsurges can even occur at the bottom of basins (Eyre). These discharges augment the inflows or can directly enter the lake (sub-lacustrine springs) and sustain the lake during its low-water phases. However, losses through infiltration are much larger in regions that are less rugged and these increase groundwater levels.

The hydrological balance of arid endorrheic lakes is a fragile balance, sensitive to any disruption, whether of a natural or human origin. This results in considerable variations in levels. Floods mostly occur in the warm season (from the end of spring to early autumn), as is the case with rivers, depending on the tropical rains, the continentality, or the melting of snows and ice in the mountains. These variations in levels, which can occur over a single season, or from one year to another (not exceeding a few decimeters), result in significant changes in volume, area and the contours of the lake, and these effects are produced more in flat lakes than deeper lakes. However, lakes in temperate deserts with cold winters and located along the middle latitudes are, on a whole, better fed and more stable than those in hot deserts at lower latitudes. Indeed, the coastlines that are represented in maps are only a statistical expression of the changing shorelines over a given period. Although they may share a common dry environment, each lake retains its own unique characteristics.

While they are water traps, endorrheic lakes are also traps for sediments. These sediments result from matter accumulating in the lacustrine basin over time, and sometimes, over several hundred meters, and which is the product of the destruction of rocks in the catchment area by agents of erosion, and transported by runoff streams and the wind. The scarcity of liquid flow in the desert, the intermittent flows, and the irregularity of floods considerably limit the transport capacities of running water concentrated in the talwegs or diffuse flows along the slopes. The transport of coarse debris (> 2 mm) is only assured in the wet seasons and in high waters, in the most powerful tributaries and intermittently, whenever floods arise. Lacustrine sedimentation in an arid zone is composed of fine formations of sand and/or silt, clays and/or salts. In the driest regions, or during the driest periods, there may also be the intercalation of sand dusts brought in by the wind. If the catchment basin is not very large and if its lithological composition is sufficiently homogeneous (small mountain lakes, for example), the sedimentation may be quite uniform. However, in lakes whose catchment basins are spread out over a large area and/or which have lithological or climatic diversity (Chad, Eyre), the sedimentation is characterized by the juxtaposition and/or the alternation of different deposits based on the season or the load carried in by the inflows. In deltas and along the coasts, the sedimentation varies depending on the nature of the littoral terrains and the competence of the rivers[3]: coarse deposits, sands and sometimes gravels or pebbles. In the medial zones, where alluvions are in transit and reshaped by the current, the local terrigenous material combines with the sediments carried in from distant sites. The finest sediments (sands, silts, clays and salts) gather in the center.

The salts arrive dissolved, having been transported in an ionic state by the runoff waters at the same time as the solid debris gathered along the slopes. Some of them are transported by the wind, after deflation along marine shores (Mauritania, Namibia), or over salinized soils in oases and agricultural lands (Turkestan). Other salts are brought in through sea spray or the rain, or through groundwater discharge (the Dead Sea, Eyre). *Salinity*, which is a measure of the dissolved salt content in lake waters, is expressed in grams per liter (g/L). In general, we only refer to the overall salinity, but it is also possible to determine the specific proportions of each component. These components differ based on the nature of the rocks in the catchment basin. They chiefly comprise of the following salts, with the dominant salt differing from one lake to another: chloride salts like halite (NaCl) and magnesium

3 For the definition of river competence, the reader can refer to section 1.4.

chlorides ($MgCl_2$) or potassium chlorides (KCl); sulfates like gypsum ($CaSO_4 \cdot 2\,H_2O$), which is as widespread as halite; carbonates like calcite ($CaCO_3$), dolomite ($CaMg(CO_3)_2$), magnesite ($MgCO_3$) or natron (Na_2CO_3). Silica (SiO_2) and aluminum (Al_2O_3) must be classified separately. They make up the composition of many rocks, such as quartzites, sandstone, igneous and metamorphic rocks, frequently found in ancient massifs and their alterites[4], sands, silts and various clay minerals. They have relatively low solubility, which varies depending on the environment. The solubility of silica depends on the crystalline state and on the pH of the medium under consideration. It increases notably when the pH exceeds nine, in an alkaline medium. Alumina is soluble in an acidic medium ($pH < 5$) and in an alkaline medium ($pH > 7$), but is insoluble in a neutral medium. The weathering of aluminosilicates, the ionic transport of the silica and alumina in rivers, the creation of neoformed clays and the deposits that are formed in the lakes are all governed by this information (Millot 1964).

In periods of high water, the salts remain dilute and widely dispersed. A large inflow of freshwater from a tributary in spate may lead to a noticeable decrease in salinity. If there is an outlet, the salts are partly removed and the salinity reduces again. On the other hand, in periods of low water, the salts become soluble in excess and change into a solid state. They are first deposited as precipitates when the saturation point is reached. Then they accumulate through concentration, due to evaporation, which leads to the sedimentation of evaporites[5], chiefly gypsum and halite. In deeper lakes with a greater volume the most soluble salts remain in solution, but the less soluble salts, like carbonates or silica, are deposited or are uptaken into the calcium or siliceous shells of living organisms. In flat lakes, like playas and sabkhas, there is a horizontal zonation of the salt content and saline sedimentation, depending on the abundance of the dissolved elements and how soluble they are. The least soluble constituents, silica, calcium or gypsum, are deposited on the foreshores in the form of crusts on the surface or sub-surface. Calcite and silica may even be conjointly deposited, as has happened in the dayas or sub-arid lakes in the north Saharan limestone hamadas, (Joly *et al.* 1954), or on the large tertiary plateaus in the Paris Basin (Bellair, Prost, and Pomerol 1962; Dewolf 1982). Other hydrated salts gather together closer to the center of the lake. These are chiefly chlorides mixed with alluvial silts and neoformed clays like kaolinite, smectite, montmorillonites, attapulgite, sepiolite and so on.

The principal factors governing the hydrological patterns and dynamics in lakes in an arid zone, are variations in temperature and the density of water, corrected by the depth of the basin.

---

4 Alterite designates a surface formation resulting from rock weathering.

5 An evaporite is a sedimentary rock formed through chemical or biochemical precipitation of a solution concentrated through evaporation.

Stimulated by the purity and dryness of the atmosphere, the air temperature is directly influenced by local variations in the solar irradiance. In temperate and cold deserts along the middle latitudes, there is a seasonal rhythm, with an absolute range that may extend over 60 or 70°C, varying between a very hot summer (>30°C) and a very cold winter (< −10°C). In hot, tropical deserts, on the other hand, seasonal variation is very small (<10°C), while diurnal variation is larger (up to >20°C). In an arid region in general, temperature has an immediate effect on evaporation, which is very high everywhere, even in the mountains and even in winter. Unlike the supply of water, which is periodic or occasional, evaporation occurs all through the year, being the chief cause for water losses in the lakes, especially small lakes. When evaporation is larger than the amount of water that has come in over a long period, especially in the case of flat and shallow lakes, it may lead to a noticeable decrease in the volume of water, a significant drop in the level and a concentration of salts. This is how the process of salinization and drying out begins. This can transform a freshwater lake into a saltwater Lake (the Holocene Saharan Lakes, for example (Petit-Maire 2002; Baumhauer, Schulz, and Pomel 2004)) and a permanent lake into a temporary lake, a daya if the water is fresh or mildly saline (Lake Chad), or a sabkha if the salinity is high (Eyre, Great Salt Lake, Dead Sea). Many arid lakes have undergone a transformation of this kind before returning to their original state, and so on.

The density of the water depends partly on the temperature and partly on the mass (or load) of particles either diluted, solid (alluvions) or dissolved (salinity). Fresh water has the property of achieving its maximum density at a temperature of 4°C and then freezing at 0°C. Usually the lighter freshwater, brackish water or warm water 'floats' on the deeper waters, which are denser as they are colder or more saline (Tiberias, Dead Sea). In an arid zone, the alluvial load is often made up of fine debris: sands, silts, clays. The arrival of this load in the lake is tied to the spring or summer floods, and the charge is significant in the case of a large flood, before it disperses and decants onto the bottom. The dissolved load, on the other hand, which is more or less concentrated by evaporation, remains suspended for a longer time and its presence increases the density of the liquid (up to 1.66 in the Dead Sea). At the same time, this dissolved load also decreases the temperature of maximum density (1.9°C for 5.8 g/L; 0.7°C for 15 g/L; 3.7°C for 35 g/L) and the freezing point, which allows certain salt lakes (the Great Salt Lake, Issyk-Kul) to remain unfrozen during winter, despite water temperatures falling below 0°C (up to −10°C in Tibet). Salinity is an essential factor in arid limnology.

Naturally, it is the surface layer of the lakes which is most affected by variations in temperature and density. This is also the layer that is most agitated by the wind, waves and currents. However, it is a relatively thin layer that rarely exceeds 30–40 m.

Beyond a depth of 40 m, the water temperature drops suddenly (thermocline)[6] and then stabilizes (homothermy) until the bottom. Because of this, the hydrological pattern of a lake depends on the climatic zone in which it is found and the depth of the basin.

– In tropical zones, the daytime temperature of the surface water can easily exceed 25°C and may even reach 40 or 50°C in some highly exposed gulfs. On the other hand, except in the high mountains (Andes) or in exceptional conditions, this temperature never drops below 4°C. In shallow lakes (depth less than 40 m; Chad, Eyre, Southern Titicaca) normal stratification, light water over heavier water, maintained by inflows of fresh water, exists in calm conditions and especially in highly saline lakes. However, the nocturnal sinking of cool water, caused by the daily thermal alternation, combined with the agitation on the surface, tends to cause a continuous blending of the waters and a total mixing affecting the entire mass of water (holomictic pattern). In deeper lakes that are cooler (over 40 m deep, in high mountains; e.g. Northern Titicaca), during the hot season only the surface layers (up to 40 m) experience mixing, with the deeper layers remaining colder and heavier. The descent of colder water only occurs in winter, setting up a homeothermic thermal profile from the surface to the bottom).

– In temperate zones, in the middle latitudes and with contrasting, seasonal thermal conditions, thermal stratification is normal in summer, with a marked thermocline at a depth of around 20 or 30 m, and homothermy in the depths. In winter, the surface temperature may fall below 4°C; the thermal stratification is then reversed with the surface, which is less than 4°C, being colder, but less dense, than water in the depths. Vertical mixing through convection takes place twice a year (dimictic pattern, double mixing): (i) in spring and (ii) in autumn, when the temperature goes beyond 4°C. The profile is then homogeneous and homothermic from the surface to the bottom. In this kind of climate, the differences between lakes are less affected by depth and more dependent on their salinity, and the effects of this salinity on the density of water and the freezing point. The septentrional part of the Caspian Sea, for example, which only has a depth of between 10 and 15 m, a salinity lower than 10 g/L and inflows of freshwater through the Volga, freezes more easily and for longer periods than the central or southern part of the Caspian Sea, which can be over 300 m deep, but where the salinity exceeds 12 g/L; the septentrional part also freezes exponentially easier than the Kara-Bogaz-Gol gulf whose salinity is over 200 g/L. Lake Balkhash, which is shallow and has low salinity, is covered with ice all through winter, while Lake Issyk-Kul, which has a minimum surface temperature of 3.8°C and a salinity of 5.8 g/L, almost never freezes. The temperature of maximum density for a salinity of 5.8 g/L is only 1.9°C, and thus vertical mixing takes place only once every winter (monomictic pattern) with total homeothermy until the deepest waters (over 600 m).

6 The thermocline is the layer of water where temperature varies rapidly.

Mixing and blending play a key role in the diffusion of dissolved gases and nutrients in the lake waters. The most significant of these dissolved gases is oxygen, O_2, essential for life. Oxygenation occurs when surface water comes into contact with the atmosphere and the oxygen is carried to the depths through successive episodes of mixing, whether irregular or periodic, which affect the lake waters in different ways:

– In shallow, holomictic lakes, the oxygen is well distributed across the homogeneous and homothermic mass, taking into account the differences along the margins and a certain lateral gradient from the banks to the center (Chad, Southern Titicaca). The medium is eutrophic, i.e. rich in nutrients (calcium, silicium, phosphates, nitrates) brought in by the tributaries and consumed by the flora and fauna, sometimes until they are fully depleted. The biggest obstacle to the diffusion of these nutrients, through an increase in density and to the development of biomass, is excess salinity. In the sabkha-lakes (Eyre, Great Salt Lake), we still find bacteria that have adapted to salinity levels close to 300 g/L, however the plankton is singularly impoverished, both in terms of the species as well as individual species, and fish disappear above a salinity of 100 g/L. Most serious of all, perhaps, is the variation in the volume of the lakes, with the accompanying variations in salinity and area that follow, especially when the variations are irregular and can imperil the very existence of the lake.

– In deep lakes, the mixing does not go all the way to the bottom layer, especially if this layer is too cold or salty, i.e. too dense to participate in the convective movement (Dead Sea). We then find a higher layer (epilimnion) where the renewing of water and nutrients can easily occur, and a lower level (hypolimnion), which is never touched, except sometimes in winter (meromictic lakes), separated by a well-marked thermocline. The depths thus have insufficient oxygen, or may sometimes lack oxygen altogether. The environment is euxinic (with reference to the Black Sea, called the Pontus Euxinus in Antiquity), unable to support any life forms except a few bacteria that have adapted to these conditions. This lack of oxygen is compensated, in a way, by the formation of hydrogen sulfide, H_2S, which increases in abundance the deeper the zone is and the longer it is confined and lacks oxygen.

Lakes in arid zones occupy a special place in the physical and anthropic water cycles (Touchart 1996). At the lowest point of endorrheic, hydroclimatic basins, they draw populations to their shore, but are fragile and hard to use. Two major obstacles threaten the survival of these ecosystems: excess salinity and fatal drying out.

2.4. References

Abrahams, A. D. and A. Parsons, eds. (1994). *Geomorphology of Desert Environments*. London: Chapman and Hall. 674 pp.

Angelot, V. d. (1843). “Recherche sur le haut degré de salure de divers lacs placés dans le fond de grandes dépressions du sol des continents, suivies de considérations sur l’origine du sel gemme en couches.” In *Bulletin de la Société Géologique de France*. 1st ser. 14, pp. 356–392.

Aubouin, J., R. Blanchet, F. Roure, and M. Tardy (1986). “Traits généraux des cordillères de l’ouest des États-Unis.” In *Bulletin de la Société Géologique de France*, pp. 741–754.

Baumhauer, R., E. Schulz, and S. Pomel (2004). “Environmental changes in the Central Sahara during the Holocene – The Mid-Holocene transition from freshwater lake into sebkha in the Segedim depression, NE Niger.” In. *Paleoecology of Quaternary Drylands*. Ed. by P. F.-H. W. Smykatz-Kloss. Vol. 102. Lecture Notes in Earth Sciences. Springer, pp. 31–45.

Bellair, P., A. Prost, and C. Pomerol (1962). “Géologie - Les hamadas tertiaires du Bassin Parisien.” In *Comptes Rendus de l’Académie des sciences* 255.19, p. 2479.

Berg, L., ed. (1932). *Lake Issyk Kul: Hydrology, Ichthyology, and Fish Production. Proc. Kirg. Special Expedition of 1932-1933*. Vol. 111, pp. 7–56.

Bouchardeau, A. and R. Lefèvre (1957). *Monographie du lac Tchad*. Vol. 1. Fort-Lamy: ORSTOM, CSLT. 112 pp.

Chevallier, P., J. Claude, B. Pouyaud, and A. Bernard (1985). *Pluies et crues au Sahel. Hydrologie de la Mare d’Oursi, Burkina Faso (1976-1981)*. Travaux et Documents 190. ORSTOM. 251 pp.

Coque, B., F. Debaine, and J. Hazemann (2000). “L’eau dans le désert du paléodelta de l’Ili (Sud du Lac Balkash, Kazakhstan): bilan d’une reconnaissance géomorphologique et apport de la télédétection (Water presence in the Ili river paleodelta desert (South of the Balkhash Lake) during the upper pleistocene and the holocene: a survey based on satellite and geomorphological data).” In *Bulletin de l’Association de géographes français* 77.1, pp. 3–9.

Currey, D. R. (1990). “Quaternary palaeolakes in the evolution of semidesert basins, with special emphasis on Lake Bonneville and the Great Basin, USA.” In *Palaeogeography, Palaeoclimatology, Palaeoecology* 76.3-4, pp. 189–214.

Debelmas, J., G. Mascle, and C. Basile (2008). *Les grandes structures géologiques*. 5th ed. Sciences sup. Paris: Dunod. 322 pp.

Dejoux, C. (1994). “Lake Titicaca.” In *Ergebnisse der Limnologie* 44, pp. 35–42.

Dejoux, C. and A. Iltis, eds. (1992). *Lake Titicaca. A Synthesis of Limnological Knowledge*. Monographiae Biologicae 68. Kluwer Academic Publishers. 576 pp.

Dewolf, Y. (1982). *Le contact Île-de-France - Basse Normandie – Évolution géodynamique*. Mémoires et documents de Géographie. Paris: Éditions du CNRS. 253 pp.

Dresch, J. (1968). “Reconnaissance dans le Lut (Iran).” In *Bulletin de l’Association de géographes français* 45.362, pp. 143–153.

— (1975). “Reflections on the future of the semi-arid regions.” In *African Environment: Special Report* 1, pp. 1–8.

Dubief, J. (1953). *Essai sur l’hydrologie superficielle au Sahara*. Algiers. 457 pp.

Erhart, H. (1953). "Sur les cuirasses termitiques fossiles dans la vallée du Niari et dans le massif du Chaillu (Moyen Congo, AEF)." In *Comptes Rendus de l'Académie des sciences* 237.6, pp. 431–433.

Fort, M. and O. Dollfus (1992). "Questions de géomorphologie dans l'ouest du Kunlun et du Tibet." In *Annales de Géographie*, pp. 371–412.

Francou, B. (1993). *Hautes montagnes: passion d'explorations*. Paris: Masson. 222 pp.

Frecaut, R. (1979). "Les bilans hydrologiques des lacs." In *Annales de Géographie* 485, pp. 1–85.

Gac, J.-Y. (1980). *Géochimie du bassin du Lac Tchad. Bilan de l'altération, de l'érosion et de la sédimentation.* Travaux et Documents 123. Paris: ORSTOM. 251 pp.

Garba, Z., A. Durand, J. Lang, and K. Alzouma (2003). "Signification paléoenvironnementale des minéraux argileux dans les dépôts de dépressions fermées du bassin du lac Tchad au Tardiglaciaire et à l'Holocène (Termit et Bougdouma, Niger oriental)." In *African Geoscience Review* 10, pp. 141–152.

George, P. (1947). *URSS, Haute-Asie, Iran.* Vol. 1. Presses universitaires de France. 535 pp.

Gilbert, G. K. (1890). *Lake Bonneville.* Vol. 1. US Geological Survey Monographs 1. Washington, DC: US Government Printing Office.

Habermehl, M. (1980). "The Great Artesian Basin - Australia." In *Journal of Australian Geology and Geophysics* 5, pp. 9–38.

Herodotus (ca. 443 BCE–420 BCE). *L'enquête*. Trans. by A. Barguet. Bibliothèque de la Pléiade. Paris: Gallimard. 1979.

Hills, E. S. (1940). "The lunette, a new land form of aeolian origin." In *The Australian Geographer* 3 (7), pp. 15–21.

Hobbs, J., J. Lindesay, and H. Bridgman (1998). *Climates of the South Continents - Present, Past and Future*. New York: John Wiley and Sons. 297 pp.

Houghton, S. G. (1994). *A trace of desert waters. The Great Basin Story*. University of Nevada Press. 296 pp.

Joly, F. (1962). *Études sur le relief du sud-est marocain.* Travaux de l'Institut scientifique chérifien, série Géographie physique 10. Rabat. 578 pp.

— (1968). "Hydrographie aride." In. *Mélanges offerts à Maurice Pardé.* Grenoble: Ophrys, pp. 323–326.

Joly, F. et al. (1954). *Les Hamadas Sud-Marocaines (Résultats de la mission d'étude, 1951, de l'Institut scientifique chérifien et du Centre de recherches sahariennes).* Mémoires de l'Institut Scientifique, Rabat.

Kotwicki, V. (1986). *Floods of Lake Eyre.* Adelaide: Engineering and Water Supply Department. 99 pp.

Létolle, R. and M. Mainguet (1993). *Les grands lacs d'Asie - L'Aral.* Paris: Science. 357 pp.

Létolle, R. and L. Touchard (1998). *Les grands lacs d'Asie*. Paris: L'Harmattan. 232 pp.

Létolle, R. and M. Mainguet (2004). "Histoire des rivières d'Asie centrale depuis deux millions d'années: certitudes et spéculations." In *Cahiers d'Asie centrale* 11/12, pp. 317–341.

Mabbutt, J. (1977). *Desert landforms. An introduction to systematic geomorphology.* Vol. 2. Canberra: Australian National University Press. 340 pp.

Maley, J. (1993). "The climatic and vegetational history of the equatorial regions of Africa during the upper Quaternary." In. *The archaeology of Africa: foods, metals and towns.* Ed. by T. Shaw, P. Sinclair, B. Andah, and A. Okpoko. London: Routledge, pp. 43–52.

Martonne, E. d. and L. Aufrère (1928). "L'extension des régions privées d'écoulement vers l'océan." In *Annales de Géographie* 37.205, pp. 1–24.

Mascle, G., B. Delcaillau, and G. Hérail (1990). "La formation de l'Himalaya." In *La Recherche* 21 (1), pp. 30–39.

Meigs, P., ed. (1977–1979). *Carte de la répartition mondiale des régions arides, Carte à 1/25 000 000, 1 feuille en couleurs, Notice explicative.* Notes Techniques Man and Biosphere 7. Paris: UNESCO. 55 pp.

Millot, G. (1964). *Géologie des argiles.* Paris: Masson. 500 pp.

Mutin, G. (2000). *L'eau dans le monde arabe.* Paris: Ellipses. 156 pp.

Neev, D. and K. O. Emery (1967). *The Dead Sea: depositional processes and environments of evaporites.* Bulletin 41. Geological Survey of Israel. 147 pp.

Nissenbaum, A. (1993). "The Dead Sea - an economic resource for 10000 years." In *Hydrobiologia* 267, pp. 127–141.

Noin, D. (1997). *L'Humanité sur la planète (carte mondiale).* Paris: UNESCO.

Olivry, J.-C., A. Chouret, G. Vuillaume, J. Lemoalle, and J.-P. Bricquet (1996). *Hydrologie du lac Tchad.* Monographies Hydrologiques 12. Paris: ORSTOM. 266 pp.

Papon, P. and L. Touchart (2003). "Balaton lake: archetypal lake-pond?" In *Annales de Géographie* 632, pp. 339–356.

Petit-Maire, N. (2002). *Sahara - Sous les sables... des lacs.* Paris: CNRS. 127 pp.

Pias, J. (1970). *Les formations sédimentaires tertiaires et quaternaires de la cuvette tchadienne et les sols qui en dérivent.* Mémoires. Paris: ORSTOM. 468 pp.

Pourriot, R. and M. Meybeck (1995). *Limnologie générale.* Paris: Masson. 956 pp.

Pouyaud, B. and J. Colombani (1989). "Les variations extrêmes du lac Tchad: l'assèchement est-il possible ?" In *Annales de Géographie* 98, pp. 1–23.

Rubrouck, G. d. (1255). *Voyage dans l'Empire mongol.* Trans. by C. Kappler and R. Kappler. Paris: Payot. 1985.

Sanlaville, P. (2000). *Le Moyen-Orient arabe. Le milieu et l'Homme.* Collection U. Paris: Armand Colin. 264 pp.

Savvaitova, K. and T. Petr (1992). "Lake Issyk-Kul, Kirgizia." In *International Journal of Salt Lake Research* 1.2, pp. 21–46. DOI: 10.1007/BF02904361.

Schneider, J.-L. (1994). *Le Tchad depuis 25000 ans: géologie, archéologie, hydrogéologie.* Elsevier Masson.

Servant, M. (1973). *Séquences continentales et variations climatiques: évolution du bassin du Tchad au Cénozoïque supérieur*. Travaux et Documents 159. Paris: ORSTOM. 573 pp.

Stansbury, H. (1852). *Exploration and survey of the valley of the Great Salt Lake of Utah*. Philadelphia, Pennsylvania: Lippicott, Grambo & Co. 457 pp.

Tilho, J. (1928). "Variations et disparition possible du Tchad." In *Annales de Géographie* 37.207, pp. 238–260.

Touchart, L. (1996). "Les lacs dans le cycle physique et anthropique - Essai de typologie." In *Bulletin de l'Association de géographes français* (2), pp. 108–118.

Tricart, J. (1954). "Une forme de relief climatique, les sebkhas." In *Revue de Géographie Physique et de Géologie Dynamique* 5, pp. 97–101.

Twidale, C. R. and J. A. Bourne (2004). "Neotectonism in Australia: its expressions and implications/Néotectonique en Australie: ses manifestations et ses implications." In *Géomorphologie: relief, processus, environnement* 10.3, pp. 179–193.

Zhao, S. and X. Xia (1984). "Evolution of the Lop desert and the Lop Nor." In *The Geographical Journal* 150 (3), pp. 311–321.

3

Hydrographic Networks

Not all hydrographic networks have left behind spectacular traces on the ground. On wide, inclined glacis and on vast plains, these traces are most often seen in the form of simple sedimentary evidence (alluvial material that is more or less separated into different classes and rounded, intersecting stratifications, lacustrine formations, etc.) or vegetation stripes, or indeed by the presence of a talweg. In the mountains and the foothills, these traces appear in the form of proper V-shaped valleys or flat-bottomed valleys, arranged in a hierarchy with confluences, flood beds and low flow channels crowded with alluvions, meanders, gorges and terraces – everything, in fact, except for running water. All that remains of a former, active hydrographic system is the stage set.

Indeed, the factors that govern the birth and development of hydrographic networks in arid environments are largely the same as those that come into play in more humid regions. The difference lies in the relative importance of the role each one plays and its power and, especially, the discontinuity of their action over space and time (Dubief 1953; Joly 1962; Rodier 1981; Mainguet 1995).

As a result, arid hydrography has certain unique features that must be highlighted.

3.1. Endorrheism–arheism

To start with, apart from coastal rivers and the large exorrheic rivers that drain into the "global ocean", all other water bodies have endorrheic drainage, i.e. they do not drain into the sea. They are either located around closed depressions of all sizes, which

Chapter written by Fernand JOLY.

may or may not contain several water bodies, or the flow dies away before the river can reach the level of the sea. The world map of arheism and endorrheism (Martonne and Aufrère 1928) coincides, for the most part, with the world map of deserts. This is primarily a climatic phenomenon: endorrheism and arheism are the direct climatic expressions of desert hydrography.

3.1.1. *Endorrheism–arheism in tropical deserts*

The deserts that are most affected are the hot, tropical deserts in Mauritania, the Thar Desert and deserts in South Africa and Australia.

In the Sahara, endorrheism–arheism begins, in the west and on the Atlantic coast itself, in the form of a coastal desert. The desert continues into the east over the entire Saharo-Arabian platform, between 15°N and 30°N, with local nuances brought in by variations in climate and relief features. The western Saharan region (Precambrian basement and sedimentary layers) is especially distinguished by superficial, linear traces that converge onto large basins (Zemmour, Djouf and Tanezrouft) that are dotted with hollows that can be flooded (playas or sebkhas) or crowded with ergs (Iguidi, Chech and Ouarane), with occasional short-lived flows, or no flows at all. The same conditions are observed in the Libyan–Egyptian desert, in the sandy corridor in central Arabia (Nefoud, Dahna, Roub al Khali) and in the Thar Desert. On the other hand, deep valleys in the foothills of the main mountains in these regions reveal the presence of proper organized and hierarchic hydrographic networks. In the northwest Sahara, many wadis arising in the subarid Atlas Mountains, oriented toward the south, run through the limestone plateaus of the hamadas, disappearing into the Grand Erg Occidental or in the Algerian–Tunisian salt lakes or *chotts*. It is only in the western Sahara that you have a powerful river, the Saoura River, that manages to flow as far as Touat; the Daoura disappears into the erg Iguidi. In the central Saharan region, around the largely volcanic relief made up of the Hoggar, Tibesti and Ennedi ranges, deep valleys diverge into neighboring basins. Downstream, on the large, inclined glacis, these valleys continue as talwegs, which are less deeply carved in the surface but still episodically and partially active. In the Sahel, the Mauritanian Adrar is surrounded by deep valleys. Further east, the Azawagh basin still sometimes receives water that comes from the Hoggar, the Adrar des Ifoghas or the *koris* from the Aïr, and which used to once flow as far as the Niger River. Still further east, the wadis arising in the Tibesti and the Ennedi Mountains have disappeared beneath the Ténéré and Borku ergs, and Chad only receives water from the tropical south (the Shari basin). A similar, radiating arrangement of deep valleys is found around the Sinai. In Arabia, rivers along the eastern slope of the crystalline ridges of the Hedjaz–Asir–Yemen region, those along the western slopes of the Oman mountain, and even those in the largely karstified limestones in Hadramaout, Dhofar and plateaus bordering the Persian Gulf, tend to get lost downstream, in ergs and the playas in the central furrow.

Like the Sahara, the tropical South African desert begins in the west on the Atlantic coast as a coastal desert: the Namib. This desert occupies the length of the coast, from the Orange River in the south till beyond the borders of Angola in the north, forming a strip of land measuring 1500 km × (30 – 100) km between the sea and the large escarpment that borders the inland Namibian plateau. It is an arid or even hyperarid desert in the sense that it almost never rains here, with the interval between rainfall sometimes stretching to several decades. At the same time, however, it is also a fog desert that is drenched in coastal fogs created by the cold water in the Benguela current and the littoral upwelling. In these conditions, this sandy desert has no surface hydrography apart from flows that disappear in the dunal hollows that can be flooded (*vleis*) and three or four exorrheic rivers, such as the Swakop or the Orange River. To the east, the aridity decreases rapidly, more rapidly in the north than in the south. The central Namibian Plateau, inclined to the southeast, is a region of subdesert aridity with inselbergs and deep valleys that are almost true canyons formed by regressive erosion around the Orange River (Fish River canyon). The Kalahari, which is called a desert as it has no permanent surface water, more closely resembles the subarid savannas in the Sahel than the real Sahara. It is an edaphic desert, where the rains (between 150 and 600 mm) are almost entirely evaporated or absorbed by a thick and permeable layer of red sands and relatively thick arboreal vegetation. In the north, the "desert" skirts the humid tropics. Here, during the Quaternary pluvials, the rivers, which were well-fed upstream by tropical rains but grew weaker in their lower courses, filled large lakes, the *pans* that are now nearly all dried out. The largest of these lakes still releases its floodwaters into the interior, marshy delta of the Okawango. The large Zambezi and Limpopo Rivers in the northeast, and the Orange River to the southwest are peripheral, and threaten the basin with exorrheism by regressive erosion.

In Australia, the desert occupies the whole center of the continent, across more than 20° of longitude and 10° of latitude. While being shorter and narrower than the Sahara, it is the largest desert in the Southern Hemisphere. Like the Sahara, it stretches between a tropical region with summer rains (in the north) and a subtropical region with winter rains (in the south). Like the Sahara, it is a flat desert (Precambrian Shield) broken up in the center by folded chains or ranges and by inselbergs (Mount Uluru and the Olga), while being depressed in the east to form a vast, subsident basin (the Great Artesian Basin). However, overall, it is a desert, being drier than the Kalahari but wetter than the Sahara, with less than 250 mm of precipitation, evaporation greater than 2,000 mm, with high interannual variability and dispersed scrub vegetation. The creeks that rise in the central mountains or peripheral relief and which can receive over 400 mm of rain are all intermittent and very irregular, and the hydrography is disrupted by almost generalized arheism–endorrheism that affects two-thirds of the continent. Like in the Sahara and in South Africa, the Australian desert extends into the west as far as the coast. The desert has less fog than the Namib, but is just as arheic, apart from a few small exorrheic rivers that come down from the littoral relief. Between the regions that receive good rainfall (between 300 and 500 mm) in the north and the arid mountains (<250 mm) in the center, there extends a desert corridor, running west to east, made up of the Great Sandy Desert and Tanami Desert. This corridor is covered by long belts of dunes (sometimes over 200 km) long, running west–east. These dunes or sand ridges are made up of red sand and are between 15 and 30 m high, parallel to each other and spaced from 300 m to 900 m apart. This corridor has no organized hydrographic network but only sporadic flows toward the closed basins (bolsons and salt lakes) that dot the plain. The same conditions are reproduced all around the central reliefs: in the Gibson Desert to the west; in the Great Victoria Desert, which has looser and more irregular sand ridges; and in the Simpson Desert in the northeast, bordering the great eastern depression, which is one of the most desertic regions in Australia with its long sand ridges running north-northwest–south-southeast (Volume 3, Chapter 1, Figure 1.43). The remainder of the depression is occupied by the hydrographic system made up of Lake Eyre, which we have already seen.

3.1.2. *Endorrheism–arheism in temperate deserts*

The phenomenon of endorrheism–arheism is also present in temperate deserts with hot summers and cold winters, as well as mountainous deserts. In the mountains, in fact, the phenomenon is helped by the morphological effects of recent tectonic activity (Meso-Cenozoic and Quaternary): an orographic factor (Martonne and Aufrère 1928) is thus added to the climatic endorrheism related to desert conditions. Aquifers, broken up by tectonic activity and dispersed into multiple syncline depressions or grabens, are usually very small but are still of interest. At the same time, given the latitude and altitude, the average temperatures are much lower and moderate evaporation, especially in winter, when the high relief leads to convection rainfall. The rivers in this region, which are better-fed through precipitation (>400 mm), snowfall and the presence of glaciers, have larger quantities of running water and often benefit from surface waters more than from groundwaters; while these flows are more powerful than the Saharan wadis, they are also intermittent and quickly die away downstream, often before reaching the lacustrine terminal base level. On the other hand, proper rivers that rise in the chief massifs and are perennial are a striking contrast to the arid plains through which they flow.

In the Middle East, beyond the large rift of Mesopotamia, drained by the Tigris and Euphrates, the desert is broken into tectonic basins embedded in the mountains of Anatolia, Iran and Afghanistan. These isolated basins, located at altitudes of over 500 m or 1,000 m, are surrounded by ridges that are between 2,000 and 4,000 m high, which receive the "Mediterranean winter rains" in the north and a few monsoon rains in the south. In northern Iran, the Dasht-e-Kavir (Great Salt Desert) is a true desert, receiving less than 100 mm of rain, with no inhabitants and dotted with temporary salt lakes that can, in rare instances, have brackish tributaries. The most arheic of these depressions is the Lut (Dresch 1968), a rift surrounded by faults, containing several hundred meters of detrital and lacustrine sediments (sands, gypsum and salt), which contain aquifers. However, surface hydrography is inadequate or wholly absent. Further east, on the other hand, in the Registan Desert (western Afghanistan), there are a series of lakes (Lake Hilmend) that are fed by water coming in from the eastern mountains and even from the Hindu Kush ranges. This arrangement of lacustrine or dried-out basins (which are the point of convergence for various flows fed by rain and springs, but which die away and disappear at the foot of the ridges) continue in the small intramountainous basins of Balochistan, south of Iran and northwest of Pakistan.

At the western end of the long, arid corridor through Central Asia is the Aral–Caspian complex (Létolle and Mainguet 1993), closed to the south and east by the Caucasus, Elburz, Hindu Kush and Pamir ranges. This complex occupies a vast depression (the Turanian depression) in a folded, peneplaned Paleozoic basement, broken up by faults and filled with Mesozoic and Tertiary sediments. Like the Chad or eastern Australia, on a continental scale, this depression is the receptacle of water that is in a doubly endorrheic catchment basin, covering over 4,000,000 km^2 (3,000,000 km^2 for the Caspian Sea and 1,500,000 km^2 for the Aral Sea), whose lowest points are the two inland "seas". The environment is a desert environment and local flows, such as the small water bodies along the shores of the lakes, are irregular and insignificant. On the other hand, the allogenous tributaries are perennial and powerful: the Volga and the Ural coming in from the north into the Caspian, and the Amu Darya and Syr Darya arising in the high mountains of central Asia and flowing into the Aral.

In eastern Kazakhstan, the Balkhash basin (500,000 km^2), despite its low altitude (340 m), is the transition point between the large Aral–Caspian complex and the basins of Upper Asia (Petrov 1962). The tributaries on the northern slopes are thin wadis, mostly dry, which disappear into the dunes and very rarely reach the lake. Those on the southern slope (including the Ili) are larger and more regular, as they have the advantage of glacier meltwater in the high mountains (up to over 4,000 m) in the south and southeast, higher rainfall (>200 mm), the presence of intermediate reservoir lakes, and less intense evaporation (1,000 mm).

The arid, mountainous regions in Upper Asia replicate and amplify Iran's hydrographic model. The hydrographic networks are cut off from one another by ridges or massifs and confined to intramountainous depressions, where altitude adds to the latitudinal position to ensure there is prevailing cold. The melting snows and glaciers trigger spring floods, while summers are scorching. The thermal differences are very large (average winter temperatures of −7°C or even −10°C and average summer temperatures greater than 25°C) and although the precipitation (largely occurring in summer) is very irregular, it is relatively abundant in the mountains (sometimes being higher than 500 mm, as opposed to less than 100 mm in the plains).

Mongolia and the Chinese province of Xinjiang have vast and deep endorrheic basins, surrounded by high, arid ridges. Most of these are ancient tectonic and lacustrine basins that have now been taken over by dunes. The Kashgar basin, drained by the Tarim River, is located at an altitude of around 1,000 m and extends over more than 1,200 km from west to east and 600 km from south to north. It lies between the mountainous block of the Pamir (>4,000 m) to the west, the sharp ridges of the Kunlun (6,000–8,000 m) to the south, and the Tian Shan (4,000–5,000 m) to the north. On its western and southern borders, at the foot of the Pamir and Kunlun ranges, which are covered in névé and small glaciers, torrential waters have constructed an alluvial piedmont of sand and pebbles, where groundwater aquifers come into contact with each other at depths ranging from 2 to 10 m. These are used for irrigation. By contrast, the southern slope of the Tian Shan, which is very arid, is only criss-crossed by dry ravines with random flows. The Taklamakan Desert (337,000 km^2) completely occupies the center of the basin and is a hyperarid desert that is totally lacking in surface flows. However, it absorbs water from the torrents that come down the Pamir and the Kunlun Mountains. The excess water spreads out and settles around the southern border of the erg in a series of lowlands that are more or less brackish (*bayirs*), covered in reeds and wild poplars (Hedin 1904). The rest of the basin is dependent on the Tarim River, subflows and deep aquifers, some of which are artesian flows that were revealed by oil drilling. In the northern part of the terminal depression of the Tarim, where Lob Nor is found (at around 750 m), the crest of the eastern Tian Shan divides into two branches that surround a deep tectonic trench, the Turfan graben (150 m below sea level). Although the ravines converging on this trench are dry, its underground resources sustain a rich oasis.

The Dzungaria opens to the north of the Tian Shan, and extends far as the Altai Mountains (4,000 m). It is a subsident basin that communicates, in the west, with the Lake Balkhash basin. The Dzungarian plain is a high plain, situated at between 200 and 600 m a.s.l., dipping to the northwest and surrounded by mountains with deep ravines. It is a flat steppe dotted with inselbergs and lacking surface flows, while being crowded with dunes and brackish lakes and marshes. Symmetrically, to the south of the Tarim depression, between two branches of the eastern Kunlun Mountains (Kunlunshan and Altunshan, 5,000 and 6,000 m a.s.l., respectively), the syncline Tsaidam basin, at an altitude of around 2500–3000 m is the point of transition with the Tibetan Plateau. This desert, made up of sand and pebbles, has few surface flows but many lakes or salty ponds (including the large Koko Nor lake, at 3,200 m) and the threat of the prevailing exorrheishm of the Huang He (Yellow River) network to the east.

The basins in the Xinjiang open in the northeast onto the Gobi Desert (or the *gobis*). "Gobi" is a Mongolian word, like the American–Mexican *bolson*, indicating a large, intramountainous endorrheic depression, dry and bare, or a steppe, with pebbly ground (*reg* in Arabic) or sandy soil (*erg* in Arabic, and *shamo* in Mongolian). Each *gobi* is divided into several secondary depressions arranged around pools or sabkhas and fed by short and poorly supplied wadis that are as irregular and unpredictable as the rains that create them. In the Chinese deserts, on either side of the upstream branch of the loop of the Yellow River, we gradually move away from the "real" deserts as there is a slight increase in pluviometry (over 150 mm) and increasingly thick steppe vegetation. However, surface flows are still absent here, or blocked by dune formations that are similar to the Saharan *aklés*, and the river, carved in loess plateaus, receives no tributary.

3.1.3. *Endorrheism–arheism in mountain deserts*

In mountain deserts, despite rainfall being higher on the whole (up to 300 or 400 mm), showers being more violent, and floods being sudden, as well as irregular, endorrheism–arheism (both orographic and climatic) is the norm. The hydrographic networks in the mountain, apart from, sometimes, the steep slopes around the peaks, are equally as threatened or degraded as on the plains.

In Tibet, the first factor is tectonic activity, which increased the number of closed depressions and disorganized the network of flows arising in the Kunlun and Himalayas. The second disruptive factor is the frequent drought, among the most severe in Asia. Each endorrheic depression has its lake or lakes, fed by the short-lived, convergent water bodies, that have few branches. This arrangement into small, isolated hydrographic units is contrasted, in the east, with the vigor of the exorrheic channels of the large rivers, the Yangtze and the Mekong.

In the basins and ranges of the temperate zone in North America, the *bolsons* are objectively equivalent to the Iranian *kavirs* and the Upper Asian *gobis*. They are subject to both summer droughts and snow and glacier retention in winter. In the Great Salt Lake Desert, to the west of Utah's Great Salt Lake, *playas* support the salt lakes or sabkhas, but tributary flows are more diffuse than hierarchic, and so much water infiltrates into the loose soil that this sometimes leads to true arheism.

In the South American Andes, within the borders of Peru, Bolivia and Chile, the high-altitude desert has all the features of a tropical climate: a small annual thermal difference, but large diurnal differences, drought and cold in the winter with random precipitations. On the high plains of the Altiplano in the Atacama, at an altitude of over 4,000 m, almost absolute desert conditions prevail, cold and bare of vegetation. Only one intermittent water body goes through this plain, the Rio Desaguadero, which only flows when Lake Titicaca overflows into Lake Poopó. The dry ravines, *quebradas*, that indent the interior cordillera (>6,000 m) show that, in times of a better climate, water flowed here, and it still sometimes flows when the rare and transient snows that form on the peaks melt. However, all traces of organized flows disappear in the plains (*punas*), and all that remains are the remnants of the salt lakes and the sabkhas (*salares*).

3.2. The disruption of hydrographic networks

Like the lakes, hydrographic networks in arid regions also used to be more extensive and more varied in their configuration in the past. Water flow systems that were well-structured and very active preceded, or alternated with, inadequate networks that were broken up or even completely abandoned. However, we are certainly seeing a largely generalized degradation of networks in most deserts, especially in the Sahara (Rognon 1989).

There are several characteristic features that reveal this degradation:

– An overall decrease in the active area of the hydrographic catchment basins. A catchment basin is both a concave topographic space delineated by the watershed mark, an impluvium that receives local precipitation (but also, like in a karst, can receive rainwater from neighboring basins through underground currents) and a drainage surface that feeds into a water body. In deserts, catchment basins that are functional even today, those that J. Dubief calls *runoff basins*, where at least from time to time some flow is seen, are noticeably smaller than the original catchment basins. A *degradation index* (Dubief 1953) may quantitatively express this decrease, in the form:

$$i = \frac{S - S'}{S} \cdot 100, \qquad [3.1]$$

where S' is the area of the runoff basin and S is the total area of the catchment basin. In the endorrheic Daoura basin (southeast Morocco), for example, out of a total area of 55,000 km^2, 25,000 km^2, or 45% of the original area, is no longer regularly in contact with the principal drain (Joly 1962), resulting in a degradation index of 55%. What remains are catchment basins that are disproportionately large compared to the wadis that drain them, a broken-up hydrography that was once alive and is now moribund, or sometimes no hydrography at all.

– A correlated decrease in the number of active drains. In the mountains and along steep slopes, or in regions with hard rocks, the networks remain branched. When the precipitation is relatively abundant (>200 mm), networks remain alive. Flow is intermittent but not absent. Links between a principal concentrated runoff drainage and its tributaries are maintained. The drainage is organized spatially, although it may be discontinuous over time. In the plains, many wadis have lost contact with the main branch and disappear before they reach it, in a string of sabkhas or alluvial cones. Each of them evolves individually. The divided networks, cut off from their weakest branches and increasingly reduced to their main drain, function as independent units guided into single, isolated channels, whose number also decreases and which are destined for total extinction. This disruption takes place gradually, from downstream to upstream, from the hyperarid regions to regions with desert aridity and then subdesert aridity, in the direction of desertification and following the deterioration of precipitation, the depletion of underground reserves and the increase in evaporation.

– Degradation of hydrological conditions, which aggravates the breakdown of networks. This is where the specific conditions of fluvial dynamics in an arid environment come into play. Poorly fed by the desultory and irregular rains and dependent on short-lived, diffuse runoff from the slopes, most of the channels are usually dry and are only revived by floods. Moving from the relatively wet regions (the peridesert borders or high relief features) toward regions that are increasingly dry, the flow dies away downstream as a result of infiltration, evaporation and a reduction in the number of inflows. This phenomenon is accelerated when humans decide to divert part of the flow, or even the entire flow, for their own purposes. Unlike temperate or tropical rivers, whose increased flow from upstream to downstream allows the water to carry an increasing solid load as far as the sea, rivers in arid regions are forced to abandon their load along the way, in the talweg itself or along interior alluvial plains, at varying distances from the starting point of the flow. In parallel to this, along the gentler slopes, diffuse, disorganized runoff divides into sheets or multiple streams that follow brief and changeable courses. Every time there is a decrease in the flow or the talweg's slope, the flood, which is weighted down to its maximum transport capacity, deposits all or part of its load, especially if the volume of water at that point is low (local floods). The single channel divides into divergent branches or is anastamosed into braided streams. Linear floods overflow and spread out (*sheet flood*), the talweg disappears and the flow infiltrates into the pores and fissures in the alluvions and the substrate. The wadi, full of its own load, can no longer flow on the surface. The deposits (the alluvial plug) are particularly large at sites that are most often reached by the strongest floods. Each of these deposits tends to become the usual terminal for a specific kind of flood, with only exceptional floods reaching the farthest deposits. The traces of wadis in arid regions, thus, are composed of a chain of alluvial plains and very flat alluvial fans, interconnected through clearly marked linear portions.

There are various causes for this degradation:

– Chief among these, of course, are climatic vicissitudes. The present-day situation shows that the breaking up of networks is increasing at the same time as endorrheism–arheism, i.e. as aridity increases. We have seen that these same geographic regions had large lakes in the Quaternary, which must have been fed by perennial water bodies or, at least, by powerful floods. For instance, Lake Bonneville, of which the present-day Great Salt Lake is but a pale reflection; the pans in the Kalahari and in

Namibia; the large Lake Eyre in Australia, and the Paleo-Chad in Africa, not to mention the mythical "inland sea", Lob Nor and the large Sahelian mesolithic lakes with their population of fisherfolk. In all these cases, we speak of the climate becoming drier, starting from the end of the Last Maximum Glaciation (20000–12000 BP) at least (Petit-Maire 2002). However, this decline was not continuous, and dry periods and wetter periods (especially during the Holocene Climatic Optimum, between 9,000 and 7,000 BP) alternated several times in the lifespan of the deserts. It must be remembered that "wetter" does not mean "very wet", as a small change was enough, either in the total quantity of precipitation or the distribution of precipitation, for denser vegetation to form and for better organized hydrographic networks.

– Other, more local, climatic causes intervene and diversify this evolution. The irregular, discontinuous rainfall, both across time and space, disrupts any interdependence between networks, as its effects across a basin are neither synchronous nor similar. Given the large intervals between rainfall and the constant evaporation, the water balance most often (though inconsistently) shows a deficit, and the lower temperatures at higher altitudes or latitudes does not make a great difference. Effective rainfall can occur anywhere and at any time, triggering a local flood at any point along the course of a flow that will then dry up for years. One stream, or section of a stream, may thus appear to be more or less degraded than another, while one lake may have more or less water than others, and some water resource may be more or less reliable than another.

– Lithology, tectonics and morphology may also contribute to the disruption of networks. The influence of the relief is indisputable. In general, along escarpments and in rugged zones, the network of ravines has undergone very little deterioration and may even be functional. On the other hand, the small incline of the piedmonts and plains, which reduce both the volume and power of the flows, leads to the water slowing down and stagnating, and, consequently, depositing its load and evaporating. We have already looked at the role of tectonic fissuration and that of hyper-permeability in karst regions in leading to deficient surface flows or even the disappearance of these flows altogether. In wadis in limestone plateaus (north Saharan hamadas, for instance), floodwaters absorbed by sediment plugs infiltrate as far as the substrate through the permeable surface formations and lead to the changes on the surface that form a doline or closed hydroeolian basin, which can be flooded (*daya*), thus prevening flow further downstream. The continuous talweg thus gets divided into a series of blind sections aligned along the old path (Joly 1962). Another significant consequence of alluvial "oversedimentation" on the plains and alluvial fans is the increase in the level of the most active drains, with respect to their neighbors, which increases the chances of capture through overflow, of changes in the direction of the path of the water flow and the consequent reorganization of the hydrographic networks. Thus, certain endorrheic rivers have been able to establish connections with the sea and become exorrheic. This is what happened with the Dra Wadi, for example, to the south of Zagora, in southern Morocco; it is also the case for the upper Niger, bending around Timbuktu, the Okawango River in the upper Zambezi, north of the Kalahari and is perhaps also the case with the Yellow River, north of Ordos.

– A final, non-negligible factor in the degradation of hydrographic networks in a dry environment is that of talwegs and alluvial plains being invaded by dunes, in the intervals between floods. The sand brought in by the wadi or by the winds settle here due to the greater humidity and denser vegetation. The dunes (barchans, nabkhas or even small ergs) cannot withstand the flow of large floods, if these occur too frequently. However, they do slow down the flow and sometimes absorb it, or are carried away as solid load, which forces the current to once again deposit the excess load somewhere, thus preparing new alluvial plains where dunes will settle again.

As Dubief (1953) showed, these different factors all result more or less directly from the aridity, which is the initial and fundamental cause, and from the temporal and spatial discontinuity in events, which is a corollary to the aridity.

3.3. Allogenous rivers and streams

Along with the networks that are poor and disorganized to varying levels in frankly desert regions, there are also large perennial rivers that border deserts, connect them, or even flow through them. These are *allogenous* rivers or streams that start in a non-arid hydroclimatic environment and which are well-fed upstream and, thus, able to transfer a renewable (and beneficial) mass of water through a region that is usually deprived of other flows. A favored, life-sustaining axis, allogenous rivers have always attracted large populations of humans (see the *World Population Density Map* at 1/15,000,000, by G. Decroix, in (Noin 1997). Some rivers have even been the starting point of large civilizations: the Nile, the Tigris, the Indus, and the Yellow River, among others.

Allogenous rivers in a desert region do not necessarily have a strong flow, but they flow continuously (toward increasingly drier regions) and their floods remain a crucial factor in fluvial flow. However, like all large rivers in an arid region, these rivers can die away downstream if they lack tributaries, as evaporation and infiltration become stronger. Their catchment basin, large and branching upstream, in the zone where water is received, is reduced downstream into a single channel that collects the flow. In general, the upstream basin benefits from greater rainfall and, sometimes, this hydrological regime compensates for, or complements, the hydrographic regime in the arid region.

Depending on the arrangement of the allogenous networks with respect to large climatic units, they can be classified into two types:

– *intrazonal* allogeny, which is located inside the same hydroclimatic system that extends over a large region;

– *interzonal* allogeny, which involves several hydroclimatic units one of which, at least, is properly arid.

3.3.1. *Intrazonal allogeny*

Intrazonal allogeny implies that there is a single hydroclimatic system. The differences between upstream and downstream here consist of nuances, rather than outright dissimilarities, and depend as much on orography as on the general atmospheric dynamics. Local precipitation is quite significant upstream, due to the effects of the relief, runoff is facilitated by the steep slopes, while the drop in temperature at high altitudes moderates evaporation and may also introduce snow and ice. Downstream, with the gentler slopes, the lower rainfall, higher temperatures, greater evaporation and the possibility of infiltration across detrital formations, the rivers lose power and load transport capacities. However, the entire basin experiences the same climate.

Intrazonal allogenous rivers can be further classified, based on the climate, into Mediterranean, temperate and tropical.

3.3.1.1. *Streams and rivers in a Mediterranean climate*

The temperatures, which are moderate on average (20–25°C), show a stark contrast between a very hot and very dry summer (>30°C) and a mild winter (10–15°C). Precipitation occurs, most notably, in the cold season, chiefly in winter and in spring, when polar or Atlantic disturbances travel through the region.

3.3.1.1.1 The Tigris and the Euphrates

The Tigris and the Euphrates (Sanlaville 2000) are two sister rivers that begin in the mountains in eastern Anatolia, at altitudes from 2,000 to 3,000 m, at the point where the west–east Taurus chain joins the northwest–southeast Zagros chain. At this point, because of the relief and the proximity to the Mediterranean and Black Sea, there is significant precipitation (from 500 to over 1500 mm), despite notable effects of continentality. The climate is semi-arid or subhumid. With this winter and the relief, the precipitation chiefly occurs in the form of snow and this sets up a pluvio-nival river regime, marked by the snows melting in April under the first spring rains. This "Mediterranean" pattern continues in the south, with the exception that the quantity of snowfall decreases the further south we move, and the quantity of rainfall also rapidly decreases, while the irregularity of the rains increases. This results in a progressive increase in subarid conditions that become properly arid in the alluvial plain of Mesopotamia.

Similar observations can be made about the wadis running down the Saharan Atlas Mountains in Algeria, like the Saoura (Vanney 1960), or those in the Moroccan Atlas

Mountains like the Daoura (Joly 1962), which flow into and disappear in the northern Sahara.

In their upper courses, the Tigris and the Euphrates share many common traits. They both arise in mountains and thus have a torrential flow and heavy alluvial loads; they have high water in spring (in April, for the Tigris, and May for the Euphrates) and very marked low flow at the end of summer (September–October). Flowing through deep, flat-bottomed valleys, they receive water from several karstic upsurges across the piedmont plateaus of the Taurus (Jezireh). The Euphrates benefits the most from this because of its tributary, the Khabur, which collects water from the Ras el Ain springs, which are among the most abundant springs in the world (Nicod 2000).

At this point, the Euphrates describes a large detour to the west, because of which its total length (2,780 km) is considerably greater than that of the Tigris (1,950 km). From here, the further they go into the arid region, the more marked the differences between the rivers become. While the Euphrates receives almost no more tributaries, the Tigris is still fed by the very irregular torrents that come down the Zagros. Consequently, the average flow of the Euphrates in Mesopotamia (775 m^3/s at Hït) is much slower than that of the Tigris (1,410 m^3/s at Bagdad)[1]. Finally, the rivers join up again, through large, diffluent valleys, with talwegs bordered by sandy-silty alluvial levées, in a marshy and saline zone that precedes their common estuary, the Shatt el-Arab, that opens into the Persian gulf, downstream of Basra.

3.3.1.2. *Streams and rivers in "temperate", continental, steppe deserts in the Northern Hemisphere, lying between 30°N and 50°N, largely in Asia*

In these regions, in addition to the climate, it is the relief, high plateaus, ridges and basins that extend to altitudes of over 4,000 m or 5,000 m that determine the hydrological differences between the various river basins. The climate is generally semi-arid or sub-arid, with stark contrasts. The average temperatures (<15°C across the year) are lower than those in Mediterranean regions. Even though the altitude moderates the summers, they are still very hot (from 25°C to over 30°C) and the winters, exacerbated by the mountains and continentality, are freezing (from −8°C to −10°C, and down to −20°C in Mongolia). Precipitation, however, is more nuanced and there is a clear transition, from the west to the east, toward aridity. The annual quantity of precipitation is low across the region (less than 500 mm), with the precipitation sometimes occurring in winter as snow (through the advection of polar air), and sometimes occurring in summer as rain (through convection or storms) west of 75°E.

1 The values for the length, the surface area, the flow rate, etc, vary depending on the author, which tributaries are taken into account, the date or reference station, etc. The numbers used here are those that are most commonly cited, or the most realistic.

In the Turanian depression and along the peaks of the Pamir, the precipitation occurs largely during the cold season, the end of autumn or the beginning of spring. A large part of the precipitation falls as snows that disappear in May–June. At the center, between the 75th and 90th meridians, precipitation occurs more rarely, and the pattern is more irregular. Cyclonic rains in the winter, which become less and less frequent, are superseded by random summer rains, in the form of squalls or showers. Finally, in the extreme east, in the Chinese and Mongolian deserts, precipitation is concentrated in the summer, with the Chinese monsoon, following a long winter drought. Similar conditions are found in the North American West.

3.3.1.2.1. The Amu Darya and Syr Darya

The Amu Darya (2,540 km) and the Syr Darya (2,212 km, or 3,019 km with its tributary, the Naryn) (Létolle and Mainguet 1993) both rise high in the Pamir and Tian Shan Mountains, at an altitude close to 5,000 m, along the Chinese borders, in a cold semi-arid region at a high altitude. The region is dotted by lakes and glaciers, with less than 500 mm of precipitation, the majority of which is snow. Both these rivers are major tributaries of the Aral Sea, in the large Turanian depression that experiences temperate desert aridity, with cold winters and less than 150 mm of rain (occurring chiefly in the winter) with droughts in the summer. The rivers are the only perennial rivers in the region because of their mountain origins. In the lower basins, both rivers have a nivo-glacial regime with low flow in winter (November to March) and floods in spring (melting snows) and/or in summer (melting ice and continental rain).

Both rivers enter the desert at a point that is very distant from their outlet into the Aral: the Amu-Darya (with a flow of 2,000 m^3/s) enters the desert around Termez, 1,300 km upstream, while the Syr Darya enters (with a flow of 730 m^3/s) when it exits Fergana, a little more than 1,000 km away. From here, both rivers are subject to the same conditions that all rivers face in arid plains. Instead of gaining more water, they lose water, since they go through regions with lower and lower rainfall, they no longer receive any regular inflows and their course is punctured by irrigation. They are only fed by floods upstream: a single nival flood toward the end of spring (May–June) in the Syr Darya, and a nival flood in spring (April–May) and floods from melting ice in summer (June–July) in the Amu Darya. While they retain high waters in the hot season, they are more irregular, suffer increased evaporation and become progressively depleted (the Syr Darya only discharges 430 m^3/s onto its delta). Along very small slopes, with slower flow, a very large river bed and rambling channels, the transport capacity and the competence of the currents decrease. The rivers still carry a large volume of material, but this is largely material of small caliber (pebbles, sand, silt), and they deposit a large quantity of alluvions. They thus slowly raise their bed above the level of the surrounding plains and, in case of a large flood, cause significant overflows and vast inundations with variable channels and changes in their path, leading to dead branches of the abating floods being abandoned. At the same time, the losses increase through larger evaporation, infiltration into the alluvion, as well as water removed by humans for irrigation (the single, large Karakum canal, downstream of Termez, removes almost 300 km^3/year from the Amu Darya and diverts it toward the Caspian). This is compounded by certain exacerbating climatic factors in the winter: the channels in the Amu Darya delta are usually covered in ice and the lower course of the Syr Darya is frozen from early-December to end of April. The ice presents an obstacle to the flow of the early spring floods, increases overflow and, when it finally melts, increases the volume of the floods.

3.3.1.2.2. The Tarim

The Tarim (Cheng 1993), which is 2,100 km long, is the only river to occupy the large, intramountainous Kashgar-Lob Nor depression, with a catchment basin extending across 350,000 km^2. With its tributary, the Yarkand, it is somewhat symmetric, with respect to the Pamir Mountains, to the Amu or Syr Darya. Its headwaters are on the eastern slopes of the "Roof of the World", the driest part (<200 mm) of this cold (average winter temperatures of $-8°C$ to $-10°C$) region with subdesert aridity at a high altitude (between 5,000 and 7,000 m) and random rainfall. The river is made up of three main branches. The northernmost arises in the Pamir and Tian Shan ranges (6,000 m) and go across the Kashgar oasis. The southernmost, and most active, branches (including the Yarkand) descend from the peaks of the Karakoram (7,000 m) at the western end of the Kunlun ranges. This upstream basin of the Tarim shares the same nivo-glacial conditions as the Amu Darya and Syr Darya basins, with high water in late spring and summer. However, the flow is more irregular than the Amu and Syr Darya, and in very cold or very dry years, the rivers may end up being half-dry at the point where the three branches come together, at an altitude of around 1,000 m, south of the Aksu oasis.

From this point onwards, the Tarim flows on over 1,200 km until it reaches the Lob Nor basin (750 m a.s.l.) through a vast corridor that is 150 km wide, between the Tian Shan ridge (4,000–5,000 m) and the northern border of the sandy Taklamakan Desert. On its left bank, it is pushed to the south by a series of dejection cones constructed at the outlet of the ravines that cut the chain. Each of these torrents supply water to an oasis (the major oases are Aksu and Kucha) before disappearing into the alluvial deposits and augmenting groundwater reserves. On its right bank, it receives no water, but discharges part of its waters under the dunes of the erg. In an arid desert climate, accentuated by deficient and irregular rainfall (<100 mm) the Tarim loses a lot of water on its journey, through evaporation (2,000–3,000 mm) and through infiltration but also, recently, through excessive pumping. Its annual mean flow is only 75 m^3/s. However, the talweg, constricted between the nested Quaternary terraces (Petrov 1962), is never completely dry and summer floods, coming in from some distant point, carry in a large quantity of fine materials, sands, silt and clays.

After the point where it also receives water from the Korla, which comes from the eastern Tian Shan ranges, the Tarim is no longer a perennial river. It disappears into the Lob Nor depression through a sort of delta formed by the two principal branches and multiple rambling and changing channels between the dunes. One of these branches skirts the eastern side of the Taklamakan erg. This branch fills the basin of the lake discovered by Prjevalski, but it has been practically abandoned since the early 20th Century. The other branch, which is still active, moves eastward, from the point where the Korla comes in, toward the basin of Sven Hedin's lake (750 m a.s.l.). At any rate, the Tarim arrives here, at the end of its course, depleted, loaded with fine alluvions and nearly dried-up.

3.3.1.2.3. The Colorado

In many respects, the Colorado River in North America has features similar to the Amu Darya and Syr Darya in Central Asia. A similar length (2,250 km, and 2,320 km if we consider the Green River tributary); similar headwaters in high mountains that are cold and snowy, in this case, the Rocky Mountains (4,300 m a.s.l.); similar paths through the arid desert region of Arizona. It is a true "Aral-like" configuration with the only difference being that the Colorado is a true river, like the Tigris or Euphrates, and, like these rivers, it terminates in a marine gulf, the Gulf of California.

In its high, semi-arid basin, the Colorado receives notable precipitation (400–500 mm), snow in the winter and rain in summer, bookended by more or less clearly marked droughts. All tributaries are perennial and have a nival regime, with high water at the end of spring (June) and low flow in winter (December–January). This hydroclimatic, two-seasons regime continues downstream along the middle and lower course of the Colorado as well, across the moderately desertic plateaus of Arizona, where the river flows at the bottom of 1,600 m deep gorges for 800 km through the famous canyon. However, the precipitation is reduced (< 150 mm), the snow contribution is minor and the effects of aridity are increasingly noticeable. The tributaries are no longer perennial flows, but become intermittent; winter rains partly compensate for the low nival waters upstream, while higher rates of evaporation and an accentuated dry summer moderate high waters in the summer. Natural conditions have also additionally been thrown into disarray by hydraulic installations set up on the river for agricultural purposes and for urban use (a succession of dams and reservoirs). Of an average annual flow of 700 m^3/s, the water removed amounts to close to 60% (400 m^3/s). With irrigation, the climatic droughts in the second half of the 20th Century and the excess evaporation in the reservoirs, the river is likely to have lost up to 15% of its flow since the 1930s.

In conclusion, the Colorado, like its Central Asian counterparts, behaves like all rivers in a region with desert aridity. Its specific flow, less than $0.5 ls^{-1} km^{-2}$, is among the slowest in its category, emphasizing the weakness of its tributary flows. Its average annual flow, which decreases as it moves downstream, is very irregular from one year to the next, and is sometimes subject to sudden and dangerous floods, triggered by heavy melting snows in the mountains, or some very concentrated rainstorm in the plains. The solid alluvial load carried by the current (turbidity) moves in a staccato manner in response to the floods in short but intense periods. Fed by the abundance of sandstones and crystalline geological formations in its path, this load is voluminous (over 70 kg/m^3) and chiefly composed of sand and silt. After it exits the canyon, south of Arizona, the Colorado travels through the most arid region in the southwest United States (accentuated desert aridity). Here, it receives another tributary, the Gila River, which is intermittent and itself punctured by irrigation. When entering Mexico, the Colorado must supply this country with 35 m^3/s for its needs as per an agreement (the 1944 United States–Mexico Treaty for Utilization of Waters of the Colorado and Tijuana Rivers and of the Rio Grande). Thus, it is almost completely dry as it completes its journey through Mexico before emptying into the Gulf of California through an estuary, which is subject to a tidal bore at high tide.

3.3.1.3. *Streams and rivers in tropical or sub-tropical hot deserts, characterized by high temperatures through the year (over 25°C on average)*

The seasonal differences in temperature are very small (less than 10°C or 15°C) while the diurnal differences are generally more accentuated, a phenomenon that intensifies as aridity increases. The rains, which easily exceed 1,500 mm in the mountains, may be totally absent in the desert. The pluvial alternations are more marked and more regular. The rains occur in summer, related to the passage of the Sun at its zenith and/or the summer arrival of marine air (the monsoon), while winter is generally very dry.

3.3.1.3.1. The Indus

The Indus, which is 3,000 km long, is a subtropical river (located between 23°30′N and 35°N), with nival influence, due to its origins in the mountain, but is essentially subject to the pluvial regime of the Indian monsoon.

The Indus descends from the high plateaus of western Tibet and first flows from southeast to northwest at the foot of the Karakoram, in a valley with a narrow and deeply eroded gorge. This sector of the mountains is covered in snows but is relatively dry as the humid winds from the south (the monsoon) find it hard to cross over the Kashmiri Himalayan ranges (over 8,000 m high). Deviated by the Hindu Kush ranges, the Indus moves southwards from this point and does not deviate further from this orientation until it reaches the sea. When it emerges from the Kashmiri mountains near Attok, between Peshawar and Islamabad, the river then flows in a bed with braided channels through an alluvial plain riddled with hills and rocky ledges that it traverses in cross valleys. On its right bank, which is mountainous but dry, it only receives thin flows, most of them intermittent, like the

Kabul River. On its left bank, five rivers converge with the Indus close to Rajanpur, bringing in water from Punjab, a semi-arid Himalayan piedmont (Lahore, 480 mm) with subdesert aridity (Multan, 174 mm), which dominates a subhumid (>500 mm) and exposed slope. Indeed, most of the basin that feeds the Indus (the runoff basin) is located upstream of this convergence. The five rivers of the Punjab supply a permanent flow to the river and the Indus receives practically no water after this. It then enters the frankly arid region between the Thar Desert to the east and the mountain deserts of Balochistan to the west (Jacobabad, 89 mm).

With a catchment basin extending over 950,000 km^2 and a high mean annual discharge (6,400 m^3/s), the Indus is a living and powerful river, subject to large variations, which can exceed 40% from 1 year to another. Its hydrological regime strictly adheres to the rhythm of the monsoon rains. As in tropical regions, the river has high water in summer (July–August) and low flow in winter (January). "This river is one of the largest in the world. It is in spate in the hot season and when the floods abate the people in this country sow their crops, like the Egyptians do after the Nile floods." (Ibn Battuta 1352). However, sudden floods may also arise in winter or spring, following rainfall in the mountain when the tail-end of a Mediterranean disturbance passes through, or snow suddenly melts, or a dam formed by boulders suddenly gives way. The effects of these floods can be felt as far as the lower course of the river.

In this hot and arid region (Hyderabad, in Sindh: 175 mm and 27°C), the Indus is only sustained by water brought in upstream. Like all desert rivers, it continuously loses water to evaporation, infiltration in sediments and the numerous hydraulic installations dating back over millenia that were the bedrock of the Sindh's brilliant civilization. This arid identity is also seen in the transit and deposit of a fine, but voluminous solid charge (500 $t/km^2/year$), which is responsible for the formation of an immense delta with an area of 8,000 km^2.

3.3.1.3.2. The Senegal

The Senegal (Michel 1973; Rochette 1974) is not, strictly speaking, a river in an arid region, but it is a good example of an allogenous river in a tropical environment in proximity to a desert. This is a short river (1,700 km) that connects a humid, mountainous region that receives over 1,000 mm of rain with a subarid plain that receives less than 250 mm that the river only skims through. However, its entire basin is located in the same tropical climatic system (Sahelian). There is only one rainy season, the summer (locally called "hivernage") while winter is the dry season with desert winds, the trade winds and the harmattan.

The Senegal arises in central Guinea, in the Fouta Djalon massif. This is a massif of peneplaned, Paleozoic sandstones, divided by faults into raised or tilted blocks that form a basement for the plateaus, dominated by encrusted tussocks. At an altitude of over 1,500 m a.s.l., the upper Senegal receives the water that filters through the fissured sandstones on the northern flank of the massif, the driest side because of its orientation. At this point, despite its altitude and the nocturnal cooling, temperatures are still high enough and constant enough for the evaporation to be noticeable throughout the year, while still not destroying the pluvial precipitation, which can exceed 1,700 mm. The tributaries and the chief artery (the Bafin) that make up the river are perennial. They traverse a region of subhumid savannas in steep valleys that are relatively embedded in the sandstone piedmont, as far as Kayes, in Mali, which still receives 736 mm of rainfall.

At Kayes, the Senegal is practically in the plains. Over the remaining 900 km it must travel there is only a 40 m drop in level. From here it follows a southeast–northwest orientation and, between Kayes and Bakel, is also fed by a perennial tributary on its left bank, the Falémé, which comes in from Fouta Djalon. On its right bank it is fed by two or three intermittent wadis from the Sahara. At Bakel (Senegal), the river regime is a typical tropical pluvial regime. The constant high temperatures ensure that evaporation is almost uniform throughout the year, the variations in flow are thus chiefly caused by pluviometry. There is high water in summer, between August and October, with a distinct maximum in September and a slight lag, with respect to the precipitation (July–August), which is chiefly due to the recharging of groundwater reserves and the growth in vegetation after the dry winter. The local floods on the plain augment the distant floods in the mountains and the mean annual discharge can exceed 2,000 m^3/s. The lowest flow is in spring and most marked between March and May, when the river's flow decreases to below 10 m^3/s, and may even be close to zero. Indeed, the 20th Century has seen several floods that are below the average, notably in 1910, 1940 and 1985 (Mainguet 1995), corresponding to a general decrease in surface flows (Sircoulon 1992; Dione 1995), a noticeable drop in groundwater levels and the development of hydraulic installations (for example, the Manantali dam regulates the flow to 300 m^3/s).

Although it is a tropical river, the Senegal still has certain features of rivers in arid regions:

– a very low specific mean annual discharge (at Kayes, it has a catchment basin that is 233,000 km^2 large but a flow of only 3.5$ls^{-1} km^{-2}$ (Guilcher 1979), which signifies mediocre flow over the entire basin, despite a large, mountainous presence upstream. However, the flow remains greater than the relative average flows of desert rivers ($<2 l s^{-1} km^{-2}$);

– a progressive decrease in flow downstream, with a large difference between the maximum floods ($>2,000 m^3/s$) and the deficient low flows ($<5.6 m^3/s$);

– alluvial loads that are largely sandy, displaced in stages, with each flood, but removed in this present case through the outlet into the ocean.

The Senegal flows in a very flared valley (up to 25 km wide), between banks of a low terrace, in an alluvial streambed, in perennial and branched channels. Bulges in the shore separate this bed from a large floodplain (floodbed) where the receding floodwaters leave behind retreating waters in numerous dead branches and microdepressions, leaving the field open for grazing and pluvial agriculture by the residents along the river.

From Bogué onwards, 280 km before it opens into the ocean, the Senegal follows a clear western orientation. It then flows practically along the border between a subarid zone in the north (with less than 250 mm of irregular rainfall) and a semi-arid zone in the south (with over 400 mm of rainfall in a tropical pattern, along the Sahelian facies). From Trarza in the north and Ferlo in the south, there are still intermittent wadis where, during floods, the river rushes in, along with strong tides (overflow lakes: Guier on the left bank and R'kiz on the right bank). Downstream of Rosso, where the Senegal arrives in a depleted state, the alluvions spread out into a kind of inland delta, blocked by a sandy barrier beach. Given its proximity to the ocean, rainfall is still high (Saint-Louis: 423 mm) and the environment remains semi-arid. The river, diverted by the sandy beach, then follows a north–south orientation and ends in a narrow estuary crowded with shifting, sandy shoals, swept by the tides and closed off by a powerful and dangerous barrier.

3.3.1.3.3. The upper Niger

The upper Niger is an allogenous river in a desert zone only from its springs in Timbuktu and for over 1,500 km. In this section of its course, the river strongly resembles the Senegal, except that it penetrates further into the arid region before turning toward the sea.

Like the Senegal, the Niger rises in Fouta Djalon, on the Guinean dorsal, at an altitude of around 1,500 m and in the same humid, tropical mountain climate, and the same geological and phytological environment. It first moves toward the northeast, in a mountainous sector that starts out in narrow, deep valleys, gathering water from many small tributaries. Augmented, through its right bank, by the Sankarani River, which is regulated by the Sélingué dam, it forms a navigable branch, despite some rapids, as far as Bamako in Mali. It then enters the Sahelian zone, which is initially semi-arid and subarid and then becomes properly arid from Diré onwards. From here, the Niger flows through a vast plain at an altitude of between 400 and 300 m, dominated in the east by the sandstone "cliffs" of Bandiagara. Upstream, at Mopti, the powerful Bani River, coming in from north of the Ivory Coast, merges with the Niger. From this point, it splits into several branches that sometimes merge, and sometimes further divide into diffluent branches that enclose marshes and small, fluctuating lakes (Courel *et al.* 1992). This is the morphological legacy of the Plio-Quaternary filling in of a subsident lacustrine basin, like in Chad, by a former inland delta. The chief divisions finally come back together around Timbuktu, in a large talweg oriented east, where the middle and lower course of the Niger begin, which are completely different.

Like the Senegal, again, the upper Niger is highly dependent on the pluviometry, with the same gap, for the same reasons, between precipitation (July–August) and flow (August–September).

The high waters occur during summer, during the "hivernage" when the sun is at its peak in the sky, following convection rainfall, monsoon tornadoes or stormy showers, which are irregular but frequent between July and November. In the mountains, where annual rainfall is between 1,200 and 2,000 mm and where the slopes are steep, the flow is perennial and torrential. At Koulikoro, shortly after Bamako, the mean interannual discharge of the Niger, over 80 years, is 1,437 m^3/s (Sircoulon 1992). The maximum flood occurs in September all across the high basin, but it comes progressively later downstream and correlatively, like all rivers in arid regions, the flow tends to decrease. The flood arrives at Koulikoro in mid-September or early October. In the plains, receiving less than 250 mm of rainfall, the perennial flow is maintained due to water brought in upstream, at the confluence of the Niger and the Bani River, and thunderstorms. At Mopti, after the Bani River merges with the Niger, the average flow is still 1,300 m^3/s, but the flood only arrives here in November. When it reaches the interior delta, it spreads out over an area of more than 20,000 km^2, which can be flooded, but which also has a high rate of evaporation (>2,000 mm). Slowed down by low slopes, the absence of tributaries, the increase in channels, soil humectation and the filling up of marshes and lakes, the Niger loses up to 48% of its volume (Rodier 1981). At Diré, close to Timbuktu, where the flood reaches in December, the average flow is no more than 1,100 m^3/s.

The low flow is in winter, in the dry season between November and June. However, this low flow is most marked in spring, reaching its minimum in May, in the upper Niger, and in June on the delta, where certain branches may dry out completely. Despite all this, a perennial flow still survives, even if it is reduced, sustained by the flow in the mountains. However, the whole of the upper Niger is indeed a river in an arid region, more than the Shari and also more so than the Senegal, as can be seen in its specific mean annual discharge, which is 3.3 $ls^{-1} km^{-2}$ over a catchment basin of 140,000 km^2. Another trait that marks this aridity is the excessive difference between the average of the high flows and the low flows, with the difference going up to 1,200–1,600 m^3/s or more. Finally, like the other rivers in the Sahel, the upper Niger has suffered significantly from the droughts in the second half of the 20th Century: its water inflow, estimated at being around 46 billion m^3 annually between 1900 and 1985, was only 37.7 billion m^3 between 1968 and 1985 (with 20.1 billion m^3 in the dry year 1984) (Sircoulon 1992), and only 24.4 billion m^3 in 2004.

The upper Niger and its delta are clearly a boon to those living along the river, offering irrigation, navigation, fishing and a source of energy. At the very least, they have made it possible to conceptualize an integrated development project, which was implemented more or less successfully by the Office du Niger, created in 1932 (Gallais 1967).

3.3.2. *Interzonal allogeny*

Interzonal allogeny implies the presence of several hydroclimatic units, at least one of which is truly arid in nature. The upstream and downstream regions of the fluviatile catchment area are marked by very different pluviometry, generally going from the more humid to the more arid. However, sometimes (e.g., along the Niger, the Huang He (Yellow River)) there is a return to more humid conditions at the end of the course. All along the river's course, the pluviometric variations modify the quantity and/or the hydrological regime of the water body, the initial conditions put in place upstream and the chronology of floods and low flows. Very large rivers and streams in deserts belong to this category.

3.3.2.1. *The middle and lower Niger*

Let us return briefly to the Niger River to consider the totality of its course (4,200 km) and its catchment area (around 1,500,000 km^2).

At Timbuktu, the Niger suddenly swings to the east and continues in a large talweg crowded with intertwining channels. About 100 km upstream of Bourem, the clear deepening of the valley testifies to the former capture of the upper Niger by a tributary of the lower Niger. The large Saharan curve of the river ends here, henceforth taking a southeast direction. It continues to flow for a long time, as far as Niamey in Niger, between Quaternary terraces and on an alluvial floodplain that becomes increasingly narrow, interrupted by islands and basins for receding floodwaters. Throughout this section, the rainfall increases from 200 mm (Gao, 234 mm) to 500 mm or 600 mm (Niamey, 490 mm), while evaporation decreases (>2,200 mm at Gao, <2,000 mm at Niamey). Until it enters the Republic of Niger, the river receives almost no tributaries, except, perhaps, fossil tributaries with unusual flow, like the outflow of the Tilemsi, near Gao, which is an outlet for the Adrar des Ifoghas. The flow, which is only fed by upstream floods from the upper Niger, diminishes constantly through evaporation, infiltration and irrigation, from 1,100 m^3/s at Diré to 925 m^3/s at Niamey.

Beyond Niamey, the Niger ceases to be a river in an arid region. Its valley is flanked by hills and small mountains and it traverses a series of rocky bars in the form of rapids, and then goes over small, intermediary alluvial plains, where the talweg widens into braided streams. The unequal amounts of water brought in – small tributaries from Burkina Faso and Benin on the right bank, even less water coming in through the unpredictable outlet of the Azaouak on the left bank, theoretically arising in the Aïr or even the Hoggar – do little to change the Niger's regime. At most, in the case of a very rainy year, there may be a peak in August–September resulting from local rains that precedes the November arrival of the Malian flood. To the southeast, the river goes on to receive, on its left bank, larger and larger tributaries, such as the Sokoto or, most importantly, the Benue, whose flow at the point of confluence is equal to that of the major river (3,000 m^3/s). This entire section, flowing through Nigeria, sees ever-increasing rainfall, from 600 mm to over 4,000 mm over the delta, and the flow thus increases to 7,000 m^3/s at its outlet. Increasingly larger local floods add to the upstream flood and the river regime is totally transformed. The upper Niger has a regime of a single flood, while here there is a regime with a double flood: one that comes in from upstream, reaching its maximum in January–February, the other of local origin, with a maximum in September. There is a clear low flow in June throughout the river.

If the Niger is a large river, the third largest in Africa after the Congo and the Nile, it owes its size to the hydroclimatic conditions at either end of its course. Its overall specific flow (51–61 $s^{-1}km^{-2}$) is considerably greater than the flow of desert rivers. However, its entry into the arid region causes it to lose a large part of its force. The Niger thus offers an example of double allogeny: intrazonal allogeny upstream in the upper Niger, which sees the force decreasing but no change in regime, and interzonal allogeny downstream in the middle and lower Niger, which see a rise in volume and a complex regime that combines the regime of local floods and upstream floods, between tropical aridity and subequatorial humidity.

3.3.2.2. *The Shari*

The Shari (1,200 km), the tributary flowing into Lake Chad, is also not a river of arid regions. It is an allogenous river, rising in the low mountains and hills of Central Africa north of Bangui, in a region of subequatorial humidity with over 1,000 mm of regular precipitation. When it emerges from these Central African highlands (1,000–1,200 m a.s.l.), it first flows northwest as far as the Chadian plains (300 m to 500 m), where it receives the essential flow of the Logona, which arises in the high Adamaoua ranges (1,900 m), on the Cameroon border. The river terminates in the middle of the Sahel (around 280 m) a.s.l., in a semi-arid region with less than 500 mm of highly uneven rainfall, offset by a constant, high rate of evaporation (2,250 mm), selective infiltration depending on the petrography of the substratum, non-negligible absorption by the savanna vegetation and intense irrigation practices. It is the chief collector of water for the lake, if not the only one, and thus it is of special interest in the world of deserts. In other aspects, it is an ordinary tropical river, like the upper courses of the Senegal and the Niger.

At N'Djamena, the Shari (Sircoulon 1992) is a perennial river with a relatively small mean annual discharge (1,115 m^3/s). Its high water levels occur in summer (June to November) during the rainy season. The floods that occur at the end of summer (the maximum in September–October) are triggered by highly concentrated and irregular (Sahelian regime) tropical showers ("tornadoes", 150–200 mm in less than 24 h) and are relatively moderate (a maximum of 1,720 m^3/s in 1955, and a maximum of 3,000 m^3/s in 2002). On the other hand, low flows in late winter and spring (April–June) that cap a dry season, which grows longer in the north, are very pronounced (minimum of 213 m^3/s in 1984) and in times of great drought may even lead to the near-complete drying out of the Shari (specific flow of 0.07$ls^{-1}\,km^{-2}$ in 1944, over a basin of 600,000 km^2) upstream of its confluence with the Logone.

The truly arid nature of the lower Shari-Logone is seen in the large interannual variation of its flow and, consequently, the water it brings to the Sahel, i.e. into Lake Chad. For example, from the early 20th Century until 1967, the Shari would annually bring in 40.4 billion m^3 of water to Lake Chad, as opposed to only 24.6 billion m^3 between 1968 and 1985 (Sircoulon 1992), which highlights the impact that the prolonged droughts in the second half of the 20th Century had on fluctuations in the level of the lake. With an overall reduction in the volume of the flow, more severe low flows and floods that are progressively smaller in intensity and fewer in number, the Shari can no longer regularly supply water to Lake Chad.

Like all large, tropical rivers, the Shari-Logone carries a considerable mass of fine debris as sands and silt. A large part of this load is deposited downstream, before arriving in the lake, in a large marshy delta and in the irrigation canals. Over time, the Shari has thus contributed to the filling in of the Chadian basin and to the episodic disappearance of the lake. Conversely, during the long, humid periods, the overflow from the floods frequently occurred and spread out widely over this flat land. During these periods, a threat arises of the Shari-Logone network being captured by the Niger, which is in close proximity and below it (Tilho 1928). One part of the floods in the upper Logone and its tributaries are already discharged to the west of Bongor, into the marshland close to a dominant, regressing valley, a tributary of the Benue that itself is a tributary of the Niger. If this threat of capture comes true, the Chad, cut off from its principal inflows, would face the risk of disappearing permanently.

3.3.2.3. *The Huang He (Yellow River)*

At first glance, the Huang He (Yellow River) seems to strongly resemble the Niger: it has a similar length (4,800 km), a similarly large catchment area (752,000 km^2), a similar origin in semi-arid mountains, a similar detour through an arid region, a similar direct relationship between rainfall and flow, a similarly large alluvial load during floods, and a similar termination into the sea through a vast delta. However, while the Niger lies between a tropical hydroclimatic system extending, latitudinally, between subequatorial humidity and desert aridity, the Huang He is more of a longitudinal extension into a temperate hydroclimatic system, going from the cold, continental semi-aridity at an altitude, to monsoon subhumidity, passing through a subarid region or a region with more or less attenuated aridity.

The Huang He arises in the Chinese province of Qinghai, at the northeast end of the Tibetan Plateau. This is a mountainous region, at a height of about 4,500 m a.s.l., situated in a region of high tectonic activity, at the point of contact between the Indian plate and the Asian plate. This plateau is dotted with lakes and narrow valleys. After a tortuous journey through the mountains, the river opens into the Lanzhou basin and travels northwards in a wide valley, punctuated by oases under cultivation, between two flat deserts, the Alashan to the west, and the Ordos to the east (Petrov 1962). When it makes contact with the southern edge of the Mongolian platform, the river is deviated to the east and then toward the south, along the meridian Taiyuan (Shanxi) massif. Is this deviation tectonic? or through capture? or a combination of both of these factors? The question remains open. But this deviation forms the Huang He's large half-loop, which recalls that of the Niger. The eastern branch, at any rate, is much narrower, more deeply embedded and less arid than the western branch. Diverted once more by the west-east obstacle made up by the Qin Mountains, the Huang He finally turns northeast, toward the sea, traversing the coastal plain whose formation it is partly responsible for.

Like the Niger, the Huang He is not a river in an arid region except in the middle section of its course. The upper, mountainous basin, upstream of Lanzhou, is already relatively dry (semi-arid), with less than 500 mm of precipitation, largely summer rains (July to September) provoked by rising storms or the last of the monsoon flows. Despite its altitude and the cold in winter, snow is rare and falls discontinuously. Evaporation is also weak, however, and apart from the low flow in winter, the flow is abundant, in direct response to rainfall, while the solid load (chiefly composed of periglacial debris) is overall quite small. The central section of the river, from Lanzhou to Baotou, made up of the western and northern arms of the loop, is more frankly arid, with less than 250 mm of rain (subarid) or even less than 150 mm (moderate desert aridity: Yinchuan, 145 mm). Rainfall always occurs in summer but in showers that are more concentrated and more violent than those that occur upstream. Evaporation is also more intense (1,000–2,000 mm), as is the infiltration through more permeable substrate. However, it is because of this journey through the desert that the river carries the constant sandy-silty alluvial load that gives its the name "Yellow River". The Alashan is the easternmost of the Central Asian intramountainous basins. Its flat surface, made up of alluvial conglomerates, lacustrine deposits and loess, is subject to strong deflation. The Alashan is a true desert, bare or with very little cover, formed of *gobis* that are simultaneously ergs with coarse elements, clayey, salty depressions, and dunes with mobile sand. The river loses a lot of its flow in this journey and erosion, especially wind erosion, is not very intense (409 t/km^2/year). At Baotou, the Huang He flows at 54% of its flow rate, with only 9% of its solid load (Xu 2003). The Ordos, isolated by the loop of the river, is a Loess plateau in its barest sense (Billard 1992). Like the Alashan, it is made up of continental, Cretacean sediment sequences, largely sandstones, capped by an uneven but thick layer (between 150–300 m) of Quaternary loess, which is more silty than sandy here. Given the wetter climate (semi-arid), with 300–500 mm of rain, the vegetation is denser and the dunes are more often fixed. The water flow in the ravines regains vigorous (Tang, Zhang, and Kong 1989). Erosion occurs more through water than wind (7,170 t/km^2/year), and the fine debris that results adds to the already high turbidity of the river (up to 30–50kg/m^3), so much so that in this section the Huang He only receives about 12.5% of its flow, but collects 56% of its load, largely composed of fine sandy-silty loess (Xu 2003).

The coastal plain and the delta are no longer in the arid region. They benefit from a temperate, subhumid climate with monsoons, 600–800 mm of rain in the summer and a flow that is sustained both by local rivers and upstream floods. In total its average annual discharge is likely to be around 3500 m^3/s, which is half that of the Niger or the Indus. However, there are large interseasonal and interannual gaps between the gigantic floods (20,000 m^3/s, 36,000 m^3/s in 1853) , the low flows that are almost a drying-up of the channel and very low specific flow (1.9 ls^{-1} km^{-2}, 4.6 at the mouth), which is yet another mark of a water course in arid regions, while the solid load is enormous (1.6 billion tonnes a year). The volume of this charge, which depends on various climatic and lithological factors the river encounters on its course (Ye 1994), regulates the power and danger posed by the "hyperconcentrated" floods that threaten the coastal plain (Xu 2003).

The alluvial deposits increase the height of the riverbed, sometimes up to 10 m above the plain, causing deadly inundations and changes in the course of the river. The floods with heavy loads are the most catastrophic. Their frequency is closely related to variations in the intensity of the summer monsoons and to the vicissitudes in a rural region that is very densely populated and intensely exploited. From the Middle Ages until the middle of the 19th Century, the river flowed into the sea to the south of the Shandong peninsula. In the 1850s, large inundations (that caused millions of deaths) caused the Huang He to change course toward the north of the peninsula, along the path that it now follows. The river is now firmly dammed, but remains unpredictable: a return to its former path (1887–1889) as well as smaller floods, as in 1969, can still cause destruction, and the delta moves seawards at a rate around 100 m per year.

The Huang He is the second largest river in China, after the Yangtze, but is undoubtedly the most formidable of all.

3.3.2.4. *The Nile*

With a drainage area (running from south to north) covering close to 3 million km^2, the Nile is an archetype of an interzonal allogenous river. Over a course of 6,700 km, it travels through all the large hydroclimatic systems in Africa one after another, from the equator to the Mediterranean. From early Antiquity, it has also been an alluring, mythical river, often deified. All those who have encountered it – Egyptians, Greeks, Romans, Arabs, Europeans – came under its spell: "The Egyptian Nile surpasses all other rivers on this Earth by the softness of its water, the length of its course and its high utility" (Ibn Battuta 1352).

Admirers of the Nile were fascinated by two significant questions (Besançon 1957; Nantet 2005):

– the first was the path and the starting of the Nile. In Antiquity, there was only vague, if not mythical, ideas about the Upper Nile, upstream of the First Cataract. Herodotus, who visited Egypt in the 5th Century BCE admitted that: "no one knows anything about the origin of the Nile, as Libya, which it traverses, is uninhabited desert country" (Herodotus ca. 443 BCE–420 BCE, vol. II, pp. 28–34). Nonetheless, he believed that the river came in from the west (probably confusing it with the Bahr el-Ghazal and the Niger) and that it was likely symmetrical, in Libya, to the Danube in Europe. In the time of Nero, in the first Century CE, the Roman military went as far as the marshes in the south, in Sudan. In the second Century CE, the Greek geographer Ptolemy described two distinct, convergent branches arising from large lakes situated in the mountains of Ethiopia, in the east, and from the equator in the south (the mythical Mountains of the Moon, which figured on all atlases until the early 20th Century). In the 14th Century CE, Ibn Khaldûn also spoke of two rivers, but divergent this time, that arose from two large lakes near the equator: "The first flows northwards and travels across Nubia (Sudan) [...]. This is the Nile of Egypt [...]. The second river turns to the west and continues flowing in this direction *(a direction opposite to that one mentioned by Herodotus, but the same confusion between the Bahr el-Ghazal, the Niger and the Senegal)* as far as the surrounding sea (Atlantic Ocean) [...]. This is the Nile of the Blacks (Nile al-Sudan)". At the end of the Middle Ages and up to the 18th Century, the privileged relationships that Europeans (Greeks and Portuguese) enjoyed with Abyssinia made the Ethiopian Nile (the Blue Nile) the chief branch of the river. In the 18th Century, interest shifted to the Sudanese Nile (White Nile) that merges with the Blue Nile at Khartoum. However, it was not until the 19th Century that tenacious explorers (Egyptian, French, German and English), more or less supported by the Pashas of Cairo, and often motivated by precolonial ideas, succeeded in solving the mystery of the source of the Nile. The discovery of Lake Nyanza, which was rebaptized Lake Victoria (Speke 1858), and then of its connection, through the Bahr el-Gebel, with the marshes of the White and Blue Nile, resulted in this lake being officially recognized as the source of the Nile. The legend of the Nile al-Sudan was abandoned and it was demonstrated that the Bahr el-Ghazal was an independent river. In reality, however, the Nile has multiple sources: the rivers that bring water into Lake Victoria. More of these are still being discovered in the 20th Century, in the mountains of Rwanda and Burundi, at altitudes of over 2,000 m. The real Nile is the White Nile, and the Blue Nile is a tributary of this;

– the second point of inquiry was the flood, which brought life into the Egyptian desert. Greeks, Romans and Arabs, who lived by the Mediterranean, where rivers were dry in summer and in spate in winter, were highly intrigued by seeing the Nile in spate in the summer. "I greatly desired to understand why the Nile swells and overflows after the summer solstice, for one hundred days, and [...] retreats and dies down, so it is only a thin flow of water during winter, until the return of the summer solstice" (Herodotus ca. 443 BCE–420 BCE, pp. 19–27). In the time of the Pharaohs, the flood would arrive in Egypt, punctually, in the middle of June, and the high waters would continue till November, reaching a maximum in September. The low flow, which could even result in the channel drying out, occurred in May. While the flood was regular, however, the volume and inundation were quite irregular, with fat years and lean years (which could sometimes be deadly, with famines). The reasons for these irregularities remained mysterious sometimes: "It is strange that the beginning of the flood occurs during the hottest period, at the time when there is low flow and other rivers dry up, while the start of the low flow [...] corresponds to the rise in water levels and floods in other rivers" (Ibn Battuta 1352). Various explanations, extraordinary, mythological or divine, were first proposed and discussed, and then researched in more scientific ways. The true reason, however, was that the flood was chiefly fed by rains in the upper basin, which occurred in the summer.

Today, the course and the regime of the Nile are well-known, measured and easily monitored by satellites (Pardé 1968; Besançon 1957; Guilcher 1979; Shahin 1985; Kerisel 1999; Sanlaville 2000; Nantet 2005). Let us look at three significant sections:

- the upper Nile, or the equatorial Nile, rises in the mountains that surround the Great Lakes in eastern Africa, a rainy region with forests and savannas within the borders of Uganda, Kenya, Tanzania, Burundi and Rwanda. An outlet of Lake Victoria, it flows into a sunken gorge, interrupted by waterfalls, rapids and lakes (Lake Kyoga, Lake Albert) and then, when it exits Lake Albert as the Bahr el-Gebel, it weaves its way over the glacis to the north of Uganda as far as the Sudanese border. The upper Nile has no features of a river in an arid region. It is part of an equatorial and subequatorial pluvial system with rains around the equinox, where the high water reaches a maximum in April that is much higher than the maximum mark in November, and a minimum level in January that is less marked than the July minimum. Despite the heavy rains (>1,000 mm or even 2,000 mm) and steep slopes, the flows remain moderate (average flow 2,000 m^3/s and specific flow 9 $l s^{-1} km^{-2}$ at the Sudan border), weighted by its journey through lakes and swamps, surfaces with high evaporation and decant basins;

- the middle Nile, or tropical Nile, has more features of aridity, but it is also more complex, with two branches that merge at Khartoum. On entering Sudan, the Bahr el-Gebel becomes a river in a semi-arid plain. It is wide, with multiple changing arms, it slows down and is encumbered by dense vegetation (reeds and papyrus) that form the vast Sudd marshes, a zone with high evaporation and infiltration, where the river loses over half its flow; a region with an average rainfall of 600 mm and which occupies the entire South Sudanese basin (>50,000 km^2). At this point, on its left bank, the Bahr el-Ghazal is augmented by the Bahr el-Arab, a weak Saharan inflow of about 800 m^3/s, and the Sobat flows in on its right bank, coming in from the mountains of Ethiopia, which adds to the river's water in the dry season. From this point on, the Bahr el-Gebel is called the White Nile (Bahr el-Abiod), and flows directly over a gentle slope (2 mm/km) toward the north, as far as Khartoum, across a flat and bare region where the pluviometry gradually decreases to less than 200 mm (Khartoum, 180 mm). In Khartoum, the river receives the Blue Nile, descending from volcanoes (>4,000 m) on the western edge of the Great African Rift. The Blue Nile's average flow is 2,500 m^3/s, but its specific flow (1.4 $ls^{-1} km^{-2}$) classifies it among the rivers of arid regions. The flood in the White Nile goes through Khartoum first, in June, followed, in July/August, by the flood in the Blue Nile. This flood is powerful enough to push the White Nile flood upstream and to supply 80% of the current flowing downstream. The Nile then engages in a series of channels with well-marked shores and bordered by vegetation, separated by waterfalls and rapids that make up the Six Cataracts. Between the Sixth and the Fifth Cataract, the Nile is again augmented, on its right shore, by the Atbara River, which is also from the mountains in the Ethiopian massif. This river is close to non-existent in winter but is in spate between late spring until October, reaching its maximum in August. From here on, the Nile flows on over 2,700 km with no further notable tributary, until it reaches the Mediterranean;

- the Egyptian Nile, or the Saharan Nile, is a true (allogenous) river flowing through desert aridity, in a region with random pluviometry that is less than 50 mm (just 2 mm at Assiout). While it is highly domesticated today, it still retains something of its wild nature, as it remains dependent on rains and upstream floods. The "tamed" Nile has regulated waters, but this water is brought in by the wild Nile at Wadi-Halfa, the point at which it enters Egypt and makes contact with Lake Nasser, restrained by the Aswan dam. The river remains a tropical one with its minimum flow in May, sustained by the White Nile, and maximum flow toward the end of August, because of the Ethiopian rivers. However, at Aswan, it is also a Saharan river: its flow is particularly modest (2,500 m^3/s) with respect to its upstream basin (specific flow: 0.9 $ls^{-1} km^{-2}$) and can tolerate significant differences (a flow of up to 8,500 m^3/s in September as opposed to only 500 m^3/s in May). The arrival of the floods that follow one another, depending on the rapidity of the flow and the distances covered, initiates a general movement of high water, which is propagated until the mouth of the river. The White Nile flood is first, in June, when it is not yet pushed back by the Blue Nile; this is then followed by the flood in the Atbara, in July, with the Blue Nile flood closely following, and overlapping with it. In August–September, the Atbara and Blue Nile supply 90% of the river's total flow. Finally, the remnants of the White Nile flood, swollen with waters from the Sobat flood, arrive in September. The waters then start retreating, a process that continues till May. From this time on, the White Nile, deprived of all tributaries on its left shore and dependent on rare showers (summer rains in the south, winter rains in the north) continues to lose its flow to evaporation, infiltration, underflow, and irrigation until in Cairo it has a mean discharge that is less than 2,000 m^3/s and a specific discharge of only 0.69 ls^{-1} km^{-2}. Conventionally, the flood thus arrives in Egypt around the middle of June and travels all along the valley with a regularity that has astonished everyone since Antiquity. This regular cycle, however, is frequently affected by significant interannual variations in volume (35 billion m^3 in 1984, 150 billion m^3 in 1988). The Aswan dam is supposed to correct this, but it is another manifestation of the inconsistency of desert flows.

Today, the Nile flows through a large, verdant valley, surrounded by yellow rock walls that are between 50 and 100 m high, cut out of the Eocene limestone sediments that extend as far as Aswan. During the Miocene, for a long time, the river flowed more to the west, along the border of the Libyan desert, between Wadi-Halfa and Assiout, in a vast space that is still populated by several large oases (Kharga, for instance) and which is a site where redevelopment is being considered today. The ancient Nile (Urnil) drained an entire hydrographic network, ending in a delta in the marine Fayoum gulf. The accentuated retreat of the sea in the late Miocene (Messinian, 7–5 Ma) caused a significant deepening, followed by a Pliocene infilling that erased all traces of this incision. The current course the river follows, further east, results from the combination of tectonic movements associated with changes in the Red Sea Rift, along with the alternating hollowing out/filling with alluvial deposits caused by the Quaternary climatic vicissitudes. Except in the case of outcrops (usually through faults) of the ancient underlying basement, the bed of the Nile is entirely carved out in the Plio-Quaternary filling, in the form of nested terraces. The most recent of these has been greatly modified by the millennia of human occupation and agricultural use.

While it is true that Egypt is a "gift of the river" (Herodotus ca. 443 BCE–420 BCE, vol. II, p. 5), it is to the alluvium brought in by the floods that it owes its prosperity. Like all rivers in arid regions, the Nile, when in spate, carries a large and colorful solid and dissolved load. Thus, at the start of July, observers would note that arrival of a "green Nile", weighed down by plant debris, microscopic algae and organic material from the swamps in the south. Then would come the "red Nile", given its color by the soils eroded from the Ethiopian volcanic massif. But throughout the year, the water would chiefly be a turbid yellow, colored by the sands and silt gathered from the Saharan crystalline basement and along the Cenozoic limestone in the valley. Every year, the entire valley, at the level of the talwegs and low terraces, would be submerged under 1–2 m of water, at 8–10 m above the low flows upstream, 6–8 m above them downstream. At these times, Egypt would look like an elongated gulf of the Aswan delta. "During the flood in the Nile [...] all Egypt becomes a sea and only the cities still emerge" (Herodotus ca. 443 BCE–420 BCE, vol. II, p. 97), Thus, every year the floodwaters would deposit *hods* (their load of fertilizing alluvium and dissolved salt) on the cultivated land, developed into irrigation and decant basins. The *hod* would become increasingly finer and more and more impermeable from upstream to downstream. Subject to evaporation and infiltration, these waters would sometimes also wreak damage, through erosion by the retreating currents and wind, and by increased salinity. With the creation of the high Aswan dam, designed as a reservoir lake (Lake Nasser 1970) for safety, the beneficial aspects of the floods could be controlled, but were also weakened: when the floodgates are closed in order to recharge the lake in winter, or in case of droughts, the trapped alluvium contributes to the formation of mudflats in the reservoir and are no longer brought to the fields.

The Nile is undoubtedly a large river, the fourth largest in the world, after the Amazon, the Congo and the Mississippi. Given its length and the area of its basin, it is a giant. But if you consider its mean annual discharge and specific flow, it is a dwarf, pushed down beyond the 20th rank on the list. In the 14th Century, Ibn Battuta compared the Nile to the Sind (Indus), which travels through the deserts in southern Pakistan and which, like the Nile, has high water in summer, is navigable and gave rise to a spectacular civilization on its shores (see section 3.3.1.3, p. 103). However, the Nile bears the strongest resemblance to the Huang He and, above all, to the Niger. All three rivers arise in high mountains with summer rains: abundant, subequatorial rains for the Nile, tropical and slightly more moderate rains for the Niger, and a highly weakened Indian monsoon in the case of the Huang He. All three then travel through a desert, with the largest section of their course being located in the desert. The Huang He travels through a desert with highly contrasting summer and winter temperatures, while the other two travel through the Sahara, which is always hot. All three are, thus, subject to a high loss of their flow through evaporation and infiltration, while also receiving, during their floods, a large load of fine, sandy-silty debris brought in by run-off streams and the wind. However, while the Huang He flows continuously, with no reservoir apart from the dam-lake in Liuchadam, downstream of Lanzhou (which has a few similarities to the Aswan dam-lake), the Niger and the Nile both contribute to the filling-in of subsident basins, Timbuktu and Sudd, respectively, through large interior deltas that are marshy and covered with vegetation, natural agents that weigh down and regulate their flow. All three ultimately terminate in vast coastal deltas that advance toward the sea and are situated in a climatic environment that is different from the rest of their river's drainage basin. This climate largely consists of an increase in precipitation, due to the influence of the (Chinese) summer monsoon over the Huang He, the abundant subequatorial rains falling on the Niger, and the weak (winter) Mediterranean rains over the Nile. However, while the Huang He and Niger see a considerable, consequent increase in flow, this increased precipitation has almost no effect on the Nile. Out of all these rivers, the Nile, which has the longest journey through a desert, which demonstrates perfect interzonality, and whose composite floods most ideally create a single flood that repeats regularly every year, is the most unique and remarkable of all rivers flowing through the arid world.

3.4. References

Besançon, J. (1957). *L'Homme et le Nil.* Collection Géographie Humaine 28. Paris: Gallimard. 385 pp.

Billard, A. (1992). "Le plateau des loess du Nord de la Chine." In *Annales de Géographie* 567 (Sept-Oct.), pp. 497–519.

Cheng, Q. (1993). *Tarim river research.* Nanjing: Hohai University Press. 246 pp.

Courel, M.-F., P. C. Chamard, C. Jacqueminet, J. Lerhun, and M. Togola (1992). "Évolution hydropluviométrique récente du delta intérieur du Niger (Mali)." In *5ème Colloque de l'Association internationale de Climatologie* (Dijon).

Dione, O. (1995). "Déficit pluviométrique et variation de l'écoulement dans le Sahel occidental/Rainfall shortage and variation in water flows in the western Sahel." In *Revue de géographie de Lyon* 70.3, pp. 199–203.

Dresch, J. (1968). "Reconnaissance dans le Lut (Iran)." In *Bulletin de l'Association de géographes français* 45.362, pp. 143–153.

Dubief, J. (1953). *Essai sur l'hydrologie superficielle au Sahara.* Algiers. 457 pp.

Gallais, J. (1967). *Le delta intérieur du Niger, étude de géographie régionale.* 2 vols. Dakar: IFAN. 621 pp.

Guilcher, A. (1979). *Précis d'hydrologie marine et continentale.* Paris: Masson. 389 pp.

Hedin, S. A. (1904). *Adventures in Tibet.* Hurst and Blackett, Limited.

Herodotus (ca. 443 BCE–420 BCE). *L'enquête.* Trans. by A. Barguet. Bibliothèque de la Pléiade Paris: Gallimard 1979

Ibn Battuta (1352). *Voyages.* Trans. by C. Defremy and B. Sanguinetti. 3 vols. Paris: François Maspero. 1982.

Joly, F. (1962). *Études sur le relief du sud-est marocain.* Travaux de l'Institut scientifique chérifien, série Géographie physique 10. Rabat. 578 pp.

Kerisel, J. (1999). *Le Nil. L'espoir et la colère. De la sagesse à la démesure.* Paris: Presses de l'École Nationale des Ponts et Chaussées.

Létolle, R. and M. Mainguet (1993). *Les grands lacs d'Asie - L'Aral.* Paris: Science. 357 pp.

Mainguet, M. (1995). *L'homme et la sécheresse.* Paris: Masson. 335 pp.

Martonne, E. d. and L. Aufrère (1928). "L'extension des régions privées d'écoulement vers l'océan." In *Annales de Géographie* 37.205, pp. 1–24.

Michel, P. (1973). *Les bassins des fleuves Sénégal et Gambie: étude géomorphologique.* 3 vols. Mémoires 63. Paris: ORSTOM. 383+571+1 vol. maps.

Nantet, B. (2005). *Histoire du Nil. Voyage au cœur des civilisations*. Paris: Éditions du Félin. 282 pp.

Nicod, J. (2000). "Ras-el-Aïn (Syrie N) et le problème des sources karstiques artésiennes aux marges et à l'intérieur des domaines arides / (Ras-el-Ain (N Syria) and the problem of the artesian karst springs at the border or inside of the arid areas)." In *Bulletin de l'Association de géographes français* 77 (2), pp. 189–199. DOI: https://doi.org/10.3406/bagf.2000.2163.

Noin, D. (1997). *L'Humanité sur la planète (carte mondiale)*. Paris: UNESCO.

Pardé, M. (1968). *Fleuves et Rivières*. 5th ed. Paris: Armand Colin. 241 pp.

Petit-Maire, N. (2002). *Sahara - Sous les sables... des lacs*. Paris: CNRS. 127 pp.

Petrov, M. (1962). "Types de déserts de l'Asie Centrale." In *Annales de Géographie* 71.384, pp. 131–155.

Rochette, C. (1974). *Monographie hydrologique du fleuve Sénégal*. Collection Monographies Hydrologiques 1. ORSTOM. 440 pp.

Rodier, J. (1981). "Caractères particuliers des régimes hydrologiques des zones arides." In. *Mélanges C.P. Péguy*. Grenoble, pp. 473–486.

Rognon, P. (1989). *Biographie d'un désert*. Paris: Plon. 347 pp.

Sanlaville, P. (2000). *Le Moyen-Orient arabe. Le milieu et l'Homme*. Collection U. Paris: Armand Colin. 264 pp.

Shahin, M. (1985). Developments in water science 21. Amsterdam: Elsevier.

Sircoulon, J. (1992). "Caractéristiques des ressources en eau de surface." In. *L'aridité, une contrainte au développement*. Paris: ORSTOM, pp. 53–68.

Tang, K., Z. Zhang, and X. Kong (1989). "A study on soil loss and soil degradation on the Loess Plateau." In. *Taming the Yellow River: Silt and Floods*. Ed. by L. M. Brush, M. G. Wolman, and B.-W. Huang. Dordrecht: Kluwer Academic Publishers, pp. 299–312.

Tilho, J. (1928). "Variations et disparition possible du Tchad." In *Annales de Géographie* 37.207, pp. 238–260.

Vanney, J. R. (1960). "Pluie et crue dans le Sahara Nord-occidental." In *Travaux de l'Institut de Recherches Sahariennes*. Monographies régionales 4, p. 114.

Xu, J. (2003). "Growth of the Yellow River delta over the past 800 years, as influenced by human activities." In *Geografiska Annaler*. A: Physical Geography 85 (1), pp. 21–30.

Ye, Q., ed. (1994). *Researches on Environmental Changes of the Lower Yellow River Basin and Laws of Water and Sediment Transportation*. Shandong Science and Technology Press. 223 pp.

4

Salts in Deserts

"[...] tous les lacs dont il sort des fleuves ne sont point salés; tandis que presque tous ceux qui reçoivent des fleuves sans qu'ils en sortent, sont imprégnés de sel. La mer Caspienne, le lac Aral, la mer Morte etc., ne doivent leur salure qu'aux sels que les fleuves y transportent, et que l'évaporation ne peut enlever."

(*[...] all lakes that have outlets to rivers are not saline; while nearly all lakes that receive river inlets without outlets are impregnated by salts. The Caspian Sea, Aral Sea, Dead Sea etc. only owe their salinity to those salts that are brought by rivers and cannot be removed by evaporation*)

(Buffon 1744, p. 183)

Climatic aridity, whether hot or cold, is not a sufficient condition for the formation of salts in deserts. The salts must still be brought in, either in dissolved form or by the wind. Further, by definition, water is scarce in a desert. Thus, there are both deserts with no salts as well as large, salt deserts. The Great Salt Lake in Utah, the *chotts* and *sabkhas* in the northern Maghreb region, the *playas* of North Mexico (see section 2.1.2) and the Uyuni *salar* in the Bolivian *altiplano* are all examples of salt deserts. Large quantities of salt are deposited here; although it has a negative impact on agriculture in the oases, the large quantity of salt is a source of wealth. The "salt caravans" are proof of this. For a long time, salt used to be the only means of preserving food like meat and fish (the Roman *garum*) and it was so important that wages were paid in salt (*salary* comes from the Latin *salarium*, derived from *sal, -alis = salt*.

The majority of dissolved salt is carried by groundwaters in aquifers, flowing at a depth that protects it from evaporation, and built up through the infiltration of rainwater

Chapter written by Guilhem BOURRIÉ.

upstream. The upstream zones (called "recharge zones" as they recharge groundwater tables or deep aquifers) may be very far from the zones where there is an upsurge of groundwater and where the salts are concentrated and deposited. The time taken for transfer from the recharge zone to the deposit zones may be thousands of years. These deep aquifers come to the surface when they cut through the topography, in a depression, e.g. due to subsidence. This is the case with the Saharan flexure, which determines the N-S alignment of the salt lakes (*chotts*) from Touggourt to Ouargla: the Melrhir, Merouane chotts, etc. (Figure 4.1).

Figure 4.1. *The N'zirah palm grove in Algeria Downstream, the palm grove gives way to the depression, called a chott, the center of which is occupied by saline soils, the sabkha (Photo G. Bourrié)*

Deep aquifers may also reach the surface as an artesian aquifer, when the impermeable layers above the aquifers (called aquitards) are fractured or artificially drilled.

Endorrheism, that is the existence of closed basins with no outlets to the sea, promotes the appearance of regions with salt zones. This is the case with the North Mexican *playas*, where small, temporary rivers from the Rockies spread out across the high plateaus. Tectonic activity plays a major role here in isolating basins through faults, flexures and subsidence and, thus, some of these basins lie below sea level. This is also the case with the Dead Sea (see section 2.2.2.1).

Allogenous rivers (see section 3.3), such as the Nile, the Euphrates, the Logona and the Shari have large floods and salts become concentrated and are deposited in the flood plains when the waters retreat (Gac 1979).

4.1. The nature of the salts

Certain salts conform to the conventional chemical definition, "a salt is a product of a reaction between a strong acid and a strong base", for example, common salt or halite (NaCl), sylvite (KCl), gypsum ($CaSO_4 \cdot 2\,H_2O$) or anhydrite ($CaSO_4$). These are neutral salts (Table 4.1). Apart from binary salts, made up of a single type of cation and a single type of anion, there are also ternary or quaternary types of salts, and some salts are only differentiated from others by the number of water molecules in their structure. An example of such salts are gypsum $CaSO_4 \cdot 2\,H_2O$ and anhydrite $CaSO_4$.

As can be seen, there is a wide variety of salts. In this chapter, we will use *salt* to refer to any mineral that is more soluble than gypsum (2 g/L). There are seven major cations and anions and Table 4.1 lists 17 salts, but this is not an exhaustive list. An aqueous solution cannot be at equilibrium with more phases than there are *independent* components: here, they are water, CO_2 and at most six solid phases. The variety of the salts obtained is thus linked to the possible evolution pathways. Indeed, as Earth evolved after its formation, its mineralogical complexity increased as the environments diversified (Robert and Bousquet 2013).

Binary salts			
Halite	NaCl	Sylvite	KCl
Nitronatrite	$NaNO_3$	Nitre	KNO_3
Bischofite	$MgCl_2 \cdot 6\,H_2O$		
Mirabilite	$Na_2SO_4 \cdot 10\,H_2O$	Thenardite	Na_2SO_4
Arcanite	K_2SO_4	Epsomite	$MgSO_4$
Kieserite	$MgSO_4 \cdot H_2O$	Hexahydrite	$MgSO_4 \cdot 6\,H_2O$
Anhydrite	$CaSO_4$	Gypsum	$CaSO_4 \cdot 2\,H_2O$
Ternary salts			
Glauberite	$Na_2Ca(SO_4)_2$	Langbeinite	$K_2Mg_2(SO_4)_3$
Carnallite	$KMgCl_3 \cdot 6\,H_2O$	Tachyhydrite	$CaMg_2Cl_6 \cdot 12\,H_2O$
Quaternary salts			
Polyhalite	$K_2Ca_2Mg(SO_4)_6 \cdot H_2O$	Kainite	$KMg(SO_4)Cl \cdot 3\,H_2O$

Table 4.1. *The major neutral salts*

Other salts contain carbonate, produced during the reaction of a strong base with a weak acid, CO_2, and thus have a basic reaction, for example, sodium carbonate (Na2CO3, natron), which is found in Lake Natron in Tanzania, and which the ancient Egyptians used in the mummification process (Table 4.2).

Conversely, other salts are formed in an acidic medium and contain distinctly acidic cations, such as aluminum or iron Fe(III). For example, alums such as jarosite and natrojarosite, felsöbányaite[1] and schwertmannite are found in media where sulfides oxidize to sulfuric acid.

Basic salts	
Natron	$Na_2CO_3 \cdot 10\,H_2O$
Nahcolite	$NaHCO_3$
Trona	$Na_2CO_3 \cdot NaHCO_3 \cdot 2\,H_2O$
Thermonatrite	$Na_2CO_3 \cdot H_2O$
Pirssonite	$Na_2Ca(CO_3)_2 \cdot 2\,H_2O$
Gaylussite	$Na_2Ca(CO_3)_2 \cdot 5\,H_2O$
Sulfuric acid forms	
Jarosite	$KFe_3(SO_4)_2(OH)_6$
Natrojarosite	$NaFe_3(SO_4)_2(OH)_6$
Schwertmannite	$Fe^{3+}_{16}O_{16}(OH)_{12}(SO_4)_2$
Felsöbányaite	$Al_4(SO_4)(OH)_{10} \cdot 4\,H_2O$

Table 4.2. *Major basic and acidic salts*

These may be natural, like mangroves, or anthropized, like residues from sulfide mines (Table 4.2). The pH here may go down to 2.

Accordingly, salts are formed when water evaporates and are mainly brought by groundwaters or rivers in deserts. The presence of water is a requisite for agriculture. Agriculture is therefore carried out in the oases, like those that abut the Saharan flexure from Touggourt to Ouargla (mentioned earlier) until the salinity becomes too high for any higher plants, except specifically adapted species, called *halophytes*, to survive. In even higher salinity, higher plants disappear and all that remains are algae and microbiota involving very few species, referred to as extremophiles. The oasis then gives way, downstream, to *chotts* and *sabkhas* (Figure 4.1).

1 Previously named basaluminite, but identified with felsöbányaite, that had been described before, and which is, thus, the preferred name.

4.2. The origin of salts and evaporite sequences

4.2.1. *Evaporites*

Sedimentary formations resulting from the evaporation of natural water, seawater or continental water are called *evaporites*. Evaporites were formed over various periods, in the Triassic, for instance, and, most recently, in the late-Miocene (during the Messinian salinity crisis, to be precise). During this Messinian crisis (7–5 Ma), the Mediterranean sea was isolated from the world ocean and its level decreased dramatically, leading to the formation of deep canyons in all rivers, such as in the Nile River (see section 3.3.2.4) and the deposition of evaporites. The opening of the strait of Gibraltar provoked an irruption of water from the Atlantic and the level of the Mediterranean sea increased very quickly (Clauzon 1979).

The salts that we find in the deserts can therefore simply be the result of the dissolution of ancient evaporites, however they are most often the result of the dissolution of continental rocks (i.e. the chemical weathering of granites, schists, basalts and volcanic glasses) upstream in the mountains, in the recharge zones, with the aqueous solutions being evaporated later downstream.

4.2.2. *Evaporite sequences*

4.2.2.1. *Ion chromatography in the landscape*

Chemical fractionation occurs during the alteration and transport of material from the mountainous upstream regions, which enjoy a pluvial climate, to the arid downstream regions. Elements are thus deposited in inverse proportion to their solubility. Tardy (1969) called it ion chromatography in landscapes (Figure 4.2). The most insoluble elements, Al Fe and Si, partially combine to precipitate as aluminum hydroxides, iron oxides and clay minerals, kaolinite, then smectites in soils. These elements remain mainly upstream, forming bauxites and ferricretes, while the most soluble elements (alkaline, alkaline earth metals, chloride, sulfate and hydrogen carbonate) are transported in solution form, along with some dissolved silica.

This pattern has been verified in the Chadian basin, from the red ferallitic soil upstream, in Central Africa, to the downstream regions where swelling clay minerals form in the flood plains of the Logona and the Shari, while salts are precipitated on the northern borders of Lake Chad (Gac 1979).

4.2.2.2. *Equilibrium and non-equilibrium in the precipitation of salts*

For the same cation, carbonates are less soluble than sulfates, which are, in turn, less soluble than chlorides. Calcium carbonate is less soluble than magnesium carbonate and also less soluble than potassium and sodium carbonates. In addition, Ca is more abundant than Mg in the rocks that belong to Goldschmidt's *sialferric series* (schists, granites, basalts) that make up most of the continental and oceanic lithosphere[2]. Thus, calcium carbonate (calcite) precipitates first, giving rise to calcretes.

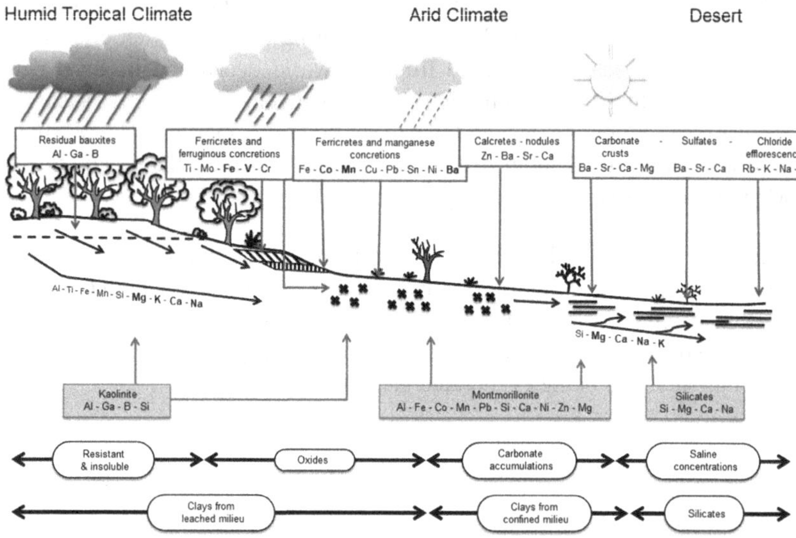

Figure 4.2. *Ion chromatography in landscapes, according to Tardy (1969), modified*

Calcite has low solubility and, for reasons related to kinetics, does not necessarily form as soon as its solubility threshold is reached. On the contrary, salts that are more soluble than calcite, such as gypsum and all other salts, are easily formed by simple precipitation when water evaporates and the solubility threshold of the salt in question is reached. In other words, salts more soluble than gypsum form at equilibrium (Bourrié *et al.* 2018).

2 Mg is more abundant than Ca in the rocks that belong to the *sifemic series* (peridotites, serpentines), which make up the Earth's mantle, and rarely forms outcrops (New Caledonia, Oman, the A Capelada massif in Galicia, Spain).

4.2.2.3. *Alkalinity and residual alkalinities*

When calcite precipitates, the concentrations of calcium and carbonate ions cannot vary independently of each other. If one of the concentrations increases, the other must decrease, such that the product of both concentrations remains constant and equal to the solubility product. Further, since carbonate is a weak base, its concentration is linked to the pH – to be more precise, its concentration increases as the pH increases. Consequently, if the pH decreases, carbonate decreases and calcium increases, whereas if the pH increases, carbonate increases and calcium decreases.

At the same time, these elements see their concentrations increasing as water continues to evaporate and the system seems to split up into two different pathways:

– in the first case, calcium increases, as does sulfate and the evolution leads to the formation of gypsum – this is the *neutral saline pathway*;

– in the second case, carbonate increases, as does sodium, and the evolution leads to the formation of sodium carbonates (trona, natron, thermonatrite) – this is the *alkaline pathway*.

Thus, after the precipitation of calcite, several possible evolutions exist, which are separated by bifurcations that depend on the initial composition of the solution (Hardie and Eugster 1970).

The problem then is that of determining how the pH varies when water evaporates. If the solution only contains neutral salts (such as NaCl, KCl, $MgCl_2$, $CaSO_4$), i.e. salts that can be obtained by mixing strong acids and bases, the pH remains constant. If some acid is present in addition to neutral salts, the pH decreases: a dilute acidic solution becomes a concentrated acidic solution, with a lower pH. Conversely, if some base is present in addition to neutral salts, the pH increases: a dilute basic solution becomes a concentrated basic solution.

Quantitatively, this is expressed as an algebraic quantity that is called the *alkalinity* of the solution. It can be calculated based on the composition of the solution or can be simply measured through the titration of the solution with a strong acid (Bourrié 1976).

Alkalinity is defined by:

$$[Alk.] = [Na^+] + [K^+] + 2[Ca^{2+}] + 2[Mg^{2+}] - [Cl^-] - 2[SO_4{}^{2-}],$$

where the values in square brackets represent the *total* concentrations of the considered elements in the solution. It is seen that for an NaCl solution, alkalinity is zero, and for an NaCl-HCl solution, it is negative ($[Na^+] < [Cl^-]$), while for an NaCl-NaOH solution, it is positive ($[Na^+] > [Cl^-]$).

The theory demonstrates that alkalinity is an increasing monotone function of pH (Bourrié 1976). The following criterion is thus applicable during evaporation *before calcite precipitates*:

$$\text{If the alkalinity is} \begin{cases} \text{positive, the pH increases;} \\ \text{zero, the pH remains constant;} \\ \text{negative, the pH decreases.} \end{cases}$$

From the precipitation of calcite, the calcium concentration is linked to pH and there is thus a relationship between these two variables and, therefore, one less degree of freedom. By removing the calcium concentration from the expression for alkalinity above, we obtain a new criterion, *calcite residual alkalinity*, whose sign can be positive, zero or negative, and which is substituted for the alkalinity with the same algebraic properties (Hardie and Eugster 1970; Al Droubi *et al.* 1980).

In practical terms, these criteria make it possible to evaluate the risk of salinization of soils during irrigation, depending on the quality of water used.

4.3. Evaporation of seawater

In coastal deserts, seawater evaporates naturally in lagoons and artificially in salt evaporation ponds. NaCl and $MgSO_4$ are the major constituents of seawater, with Ca^{2+}, $CO_3{}^{2-}$ and K^+ being minor elements. Alkalinity is slightly positive due to the presence of carbonate. Salts are precipitated in reverse order of their solubility: the first to precipitate is calcite $CaCO_3$, followed by gypsum and then anhydrite and halite. The subsequent changes are more complex and depend on the presence of inflows of continental freshwater or hydrothermal inflows (Spencer *et al.* 1990). These subsequent changes occur through the precipitation of calcium sulfate and sodium sulfate (glauberite), followed by magnesium sulfate and potassium sulfate (polyhalite, kainite, kieserite) and, finally, potassium chloride and magnesium chloride.

To explain the considerable thickness of the evaporite sedimentary series, we must consider a lagoon basin separated from the sea, opened through a bulge and periodically invaded by new sea water input (Robert and Bousquet 2013, pp. 494–498). Each cycle begins with organic matter, formed in stagnant water when the basin is closed, and then continues with the formation of calcium carbonate, anhydrite and halite, and ends with salts of potassium and magnesium. Most often the cycles are incomplete as the sea invades the lagoon before the cycle can finish.

4.4. Evaporation of continental freshwater

The evolution of freshwater depends on the initial composition of the solution and, more specifically, on the sign and magnitude of its alkalinity, followed by the sign and quantities of different residual alkalinities.

4.4.1. *The neutral saline path*

In the neutral salt path, alkalinity is mildly positive and during concentration through evaporation, the pH increases and, therefore, so do carbonate and calcium. When calcium carbonate precipitates, the residual alkalinity becomes negative to such an extent that the pH decreases, after having increased. As a result, the overall pH remains close to neutrality. When $CaCO_3$ precipitates, CO_3^{2-} decreases and Ca^{2+} increases. In parallel to this, the concentration of SO_4^{2-} increases. Thus, gypsum or anhydrite will be precipitated next. If there is any silica in the solution, magnesian clay minerals may form. While still remaining neutral, the solutions then evolve toward halite and potassium and magnesium salts, as discussed above.

This is the geochemical pathway that dominates the Sahara, in *chotts* and *sabkhas* all across North Africa. Deep groundwaters that rise through artesian activity and overflow into Chott el Djerid in Tunisia follow this pathway too (Gueddari 1983; Gueddari 1984). Consequently, the pH remains close to neutral and agriculture is possible even when irrigation is carried out using waters with salinity of up to 6 or 7 g/L.

A similar evolution is observed in the Salar de Uyuni in Bolivia.

4.4.2. *The alkaline path*

In the alkaline path, the initial alkalinity is strongly positive, and the residual calcite alkalinity remains positive. When $CaCO_3$ precipitates, the pH continues to increase, and therefore, CO_3^{2-} increases and Ca^{2+} decreases. As a result, gypsum is not formed. Simultaneously, the Mg^{2+} and silica concentrations increase and magnesian clay minerals, such as palygorskite[3], are formed. This is known as the basic chemical sedimentation (Trauth 1977). In parallel, the concentration of Na^+ increases and the solution evolves toward the formation of sodium carbonate $Na_2CO_3 \cdot 10\,H_2O$, or *natron*. This pathway is followed when the solution initially has an alkaline pH of 8 9 or 10, and salinity of between 3 and 25 g/L, as a result of

3 Palygorskite is also known as attapulgite and is a fibrous clay, with shampoo and degreasing properties, making up Fuller's earth, for example.

volcano-sedimentary environments in the lakes of the African rift (Gueddari 1984). During concentration by evaporation, the pH increases up to 11. At these pH values, clay minerals are no longer formed, instead zeolites are formed, like magadiite, for instance, which is a sodium silicate found in Lake Magadi in Kenya, or amorphous gels. The basic salts found are trona $Na_2CO_3 \cdot NaHCO_3 \cdot 2\,H_2O$ and thermonatrite ($Na_2CO_3 \cdot H_2O$). The neutral salt formed is halite. The basic salts in lake Natron (Tanzania) were used for mummification in ancient Egypt as mentioned earlier.

When the solution is initially rich in sulfate, we see the formation of sodium sulfates, mirabilite and thenardite, for example, in Deep Springs Lake in California (Gueddari 1984). Mirabilite and thenardite are also present in the saline lagoons of Aragon (Spain) (Auque *et al.* 1994).

4.4.3. *The acidic path*

In order for the pH to decrease when water is concentrated by evaporation, alkalinity must be negative. This is rarely the case, however, as the chemical weathering of rocks through hydrolysis largely liberates cations, thus consuming H^+, which leads to a positive alkalinity (Bourrié 1976). The reactions that produce a negative alkalinity are essentially the oxidation of forms reduced of N and especially of S. Sulfur, in its reduced state, forms sulfides with iron (pyrite, marcassite), arsenic (arsenopyrite), lead (galena), zinc (zinc sulfide/blende), mercury (cinnabar), etc. Sulfur is naturally present in reduced state in the Earth's mantle and lithosphere. Sulfides are also formed by the reduction of sulfates in seawater by bacteria, during the oxidation of organic material in confined environments such as as mangroves, for instance. The reoxidation of sulfides during the drying out of mangroves, or when they are drained, produces sulfuric acid.

Thus, the acidic path is largely a sulfuric acid path. At an acidic pH, aluminum is soluble and a double sulfate of aluminum and potassium is formed, jarosite or its homologue with sodium, natrojarosite. Other, rarer salts are found associated with the yellow-colored jarosite, such as the salt tamarugite $NaAl(SO_4)_2 \cdot 6\,H_2O$.

The presence of metalloids such as arsenic, and heavy metals such as lead, zinc and mercury, associated with sulfides, or aluminum dissolved by the sulfuric acid that is liberated, makes the water unfit for consumption. It is therefore important to avoid provoking the reoxidation of sulfides, a process that is both chemical and biochemical, partly autocatalytic and difficult to control once it has started. These processes are also encountered in other contexts such as water drained from sulfide and lignite mines, or sediments in the Gangetic Delta in Bangladesh, posing serious health problems.

4.5. Systems mixing surface water and deep water

4.5.1. *The Saharan system*

Once covered by sediments, ancient marine evaporites may rise to the surface because of their low density, forming "diapirs". In the Sahara, the salt deposits in Taoudeni were formed in this way. This is also the case with the "salt mountains" in Djelfa.

On the other hand, water in deep aquifers in the Continental Intercalaire and the Complexe Terminal are not saline initially. However, as they rise to the surface through artesian activity, they travel through a series of evaporites from the Upper Cretaceous which they partially dissolve. On the surface, the water becomes concentrated through evaporation and mixes with the saline surface water formed when the salt deposits in the sabkhas are dissolved in rainwater (Slimani *et al.* 2017; Bourrié *et al.* 2018).

4.5.2. *Potash deposits*

The above processes do not explain the formation of potash deposits: only small quantities of potassium minerals are formed. The magnesium and silica concentrations must be low for potassium to bond with chlorides and sulfates, instead of bonding with the magnesium and being diluted in clay minerals.

In specific tectonic conditions that lead to faults, meteoric water that has circulated in deep aquifers rises to the surface, mixes with surface water and becomes concentrated in endorrheic basins in arid conditions. This is what happens in China, for example, in the Quaidam basin, with an area of 120,000 km^2, lying to the north of the Qinghai-Tibet plateau, at an average altitude of 2,800 m a.s.l. The rivers that descend from the Kunlun mountains spread out over the Qarhan salt plains. The springs that rise up through the faults are rich in calcium and chloride; the calcium precipitates as calcite, gypsum and anhydrite. There is therefore, no calcium left to combine with magnesium to form polyhalite and the precipitation of magnesium sulfates are delayed, which allows the formation of potassium salts (Spencer *et al.* 1990).

In the absence of dissolved silica, the potassium cannot enter into the silicate networks in the magnesian clay minerals and is precipitated in the form of carnallite, $KMgCl_3 \cdot 6\,H_2O$. The determining factors here are complex: tectonic, hydrogeological, climatic and morphological (altitude), leading to the scarcity in the potassium deposits that are so important for fertilization.

4.6. Atmospheric origin of nitrate deposits

The Atacama desert in Chile, the most arid desert in the world, is located along a western ocean border and its aridity is related to continental trade winds (see Volume 1, section 3.2.2.2). Aridity set in from the middle of the Miocene epoch (from 15–10 Ma). Deposits of nitrates are neither related to the topography of the substrate nor its specific nature (tuffs, rhyolites) formed by the andesitic volcanic activity. These field observations point to an atmospheric origin of nitrates, either through the direct, dry deposition of nitric nitrogen or ammoniacal nitrogen, or an oceanic origin in the form of aerosols, brought in through fogs rising from the ocean, whose waters are rich in nitrates. Indeed, the continental trade winds, from the NE or the SE, push warm water out into the open seas, causing cold water to rise (upwelling currents) that is rich in nitrates and phosphates and that feeds the phytoplankton and, from there, the entire food chain, ensuring waters are rich in fish, thereby, attracting seabirds. In these particular ecosystems, the ocean supplies nitrogen and phosphorous to the continent. The extreme aridity of the region prevents plants from establishing themselves here, but there are algae that can fix atmospheric nitrogen and that contribute to the nitrogen input.

Nesting seabird colonies along the seashore produce accumulations of excrement, *guano*, that is rich in phosphates, such as ammonium magnesium phosphates like struvite and stercorite[4] and rich in sodium and potassium nitrates (saltpeter or niter). In the Atacama desert, however, nitrates are not systematically associated with phosphates, i.e. the guano is only a marginal source of nitrates. Instead, most nitrates are the result of oxidation of atmospheric nitrogen by ozone. This has been confirmed by analyzing stable nitrogen and oxygen isotopes (Böhlke *et al.* 1997). Another source of nitrogen is andesitic volcanic activity, which throws up NH_3 which then oxidizes to form nitrate. The presence of iodate and perchlorate, which are highly oxidized forms, is ascribed to the photochemical oxidation of chloride and iodine brought in by sea spray in the upper atmosphere (Ericksen 1981).

In order to obtain the quantity of nitrates seen in these deposits, these must have accumulated over a period of 100,000 years or even 10 million years. (Ericksen 1981; Böhlke *et al.* 1997).

If the major part of the nitrogen is brought in by the atmosphere, this means that over a period of several million years, the quantity of nitrogen brought in through volcanic activity must have been negligible or can be discounted; it follows therefore that *because of its hyperarid conditions, the Atacama desert has recorded the quantities of nitrogen and perchlorate produced naturally in the atmosphere and*

4 The name stercorite comes from the Latin *stercus* = dung.

deposited on the ground. In other words, this atmospheric deposition is likely to have been the same over the rest of the planet, and over this duration anthropic contributions were negligible too. But in regions with plants, atmospheric nitrates would have been consumed and perchlorates reduced to chlorides through reactions with reducing agents such as organic compounds and Fe^{2+}.

4.7. Mankind and salts

As can be seen, salts formed in the deserts are not limited to common salt (NaCl or halite). From earliest Antiquity, humans have used salts, mainly halite but also others such as natron $Na_2CO_3 \cdot 10\,H_2O$ and niter KNO_3[5].

Humans and animals need salt for maintaining equilibrium in their bodies and for their nervous systems (sodium pumps). Thus, we have the regular wanderings of the nomads, with their animals, looking not only for grazing land and water but also salts (salt licks). In addition, salting made it possible to preserve food, for instance the Roman *garum*, fish preserved in brine. Salt is thus a precious and rare commodity that can even be used as currency. Salary comes from the Latin *sal* (as does salad).

Table salt, halite, may be directly obtained from evaporites (Figure 4.3).

Until very recently, salt was transported across the Sahara in caravans made up of camels, which would follow regular routes, passing through a number of "desert ports", established at least as far back as the Abbassid dynasty (see Volume 1, section 2.8.1.2, "Arab rule" and section 2.9.2.2, "The salt route"); salt is the only resource traded in the Sahara itself (section 2.9.2.2 salt was exchanged for gold from Guinean mines. The salt caravan is called the *Azalai* (see Marc Côte's section dedicated to salt in the Sahara, Volume 3).

Desert populations have thus used extracted salts from marine and continental evaporites. According to certain accounts, salt and gold were considered equal in value.

Modern economies, in Europe and then the rest of the world, have been based upon a spectacular increase in agricultural yields, which required large amounts of phosphate, potassium and nitrate, resources that are very unequally distributed. Salt deposits in the deserts, either present-day or fossil deposits, were thus widely mined in the past, and are sometimes still mined today. The transport of nitrates from Chile,

5 Natron, nitrogen and niter share the same etymology. Before the advent of chemistry and the discovery of elements, the properties of sodium nitrate were not clearly attributed to nitrate, rather than to sodium. Hence, the symbols Na (natrium) and N (nitrogen) are used.

via Cape Horn, was a source of considerable prosperity for sailors and residents of Atacama, but it also provoked the Pacific war (1879–1881) between Chile and Bolivia, denying Bolivia access to the ocean. The artificial synthesis of ammonia (Born–Haber process) put an end to this trade.

Figure 4.3. *Salt fields in the Sahara, Tegguiddan Tessoum, Niger. Salt is traditionally gathered by women. Left; In the background, camels are loaded with salt (Photo Y. Dewolf)*

The presence of perchlorate in Chilean nitrates is partly responsible for the perchlorate contamination of groundwaters in northeast France, although the majority of this contamination came from WW1 explosives – those that exploded and, even more, those that remain unexploded. "Natural" fertilizers may be more polluting than "chemical" fertilizers, which are more purified.

Today, trace elements are being extracted from salts: for example lithium, used in batteries, can be extracted in the Atacama desert. However, this requires pumping brackish waters and will lead to overexploitation of the already rare water resources, to the detriment of the local Amerindian population.

Figure 4.4. *Camels being loaded to transport salt, Tegguiddan Tessoum, Niger (Photo Y. Dewolf)*

4.8. References

Al Droubi, A., B. Fritz, J.-Y. Gac, and Y. Tardy (1980). "Generalized residual alkalinity concept; application to prediction of the chemical evolution of natural waters by evaporation." In *American Journal of Science* 280, pp. 560–572.

Auque, L., V. Vallès, H. Zouggari, P. Lopez, and G. Bourrié (1994). "Importancia de la variación de solubilidad de la mirabilita con la temperatura en la evolución geoquímica de las lagunas de Los Monegros (Zaragoza)." In *Revista de la Academia de Ciencias, Zaragoza* 49, pp. 177–189.

Böhlke, J., G. Ericksen, and K. Revesz (1997). "Stable isotope evidence for an atmospheric origin of desert nitrate deposits in northern Chile and southern California, U.S.A." In *Chemical Geology* 136, pp. 135–152.

Bourrié, G. (1976). "Relation entre le pH, l'alcalinité, le pouvoir tampon et les équilibres de CO_2 dans les eaux naturelles." In *Science du Sol* 3, pp. 141–159.

Bourrié, G. et al. (2018). "Irrigation water and soil quality." In. *Soils as a Key Component of the Critical Zone 4.* Ed. by G. Bourrié. Soils Set coordinated by C. Valentin. Earth System - Environmental Sciences Series ed. by A.Mariotti London: ISTE Ltd and Wiley, New York, pp. 73-126.

Buffon, G. H. L. d. (1744). *Matières générales - I. — Théorie de la Terre.* Paris. 1848.

Clauzon, G. (1979). "Le canyon messinien de la Durance (Provence, France): une preuve paléogéographique du bassin profond de dessiccation." In *Palaeogeography, Palaeoclimatology, Palaeoecology* 29, pp. 15–40.

Ericksen, G. E. (1981). *Geology and origin of the Chilean nitrate deposits. A discussion of the characteristic features of the nitrate deposits and the sources and modes of accumulation of their saline components.* U.S. Geological Survey Professional Paper 1188. Washington, D.C.: United States Government Printing Office. 42 pp.

Gac, J.-Y. (1979). "Géochimie du bassin du Lac Tchad - Bilan de l'altération, de l'érosion et de la sédimentation." Doctoral thesis in sciences. Université Louis Pasteur, Strasbourg. 249 pp.

Gueddari, M. (1983). "Geochemistry of brines of the Chott El Jerid in southern-Tunisia - Application of Pitzer's equations." In *Chemical Geology* 39, pp. 165–178.

— (1984). *Géochimie et thermodynamique des évaporites continentales - Étude du lac Natron en Tanzanie et du Chott El Jerid en Tunisie.* Sciences Géologiques, Mémoire 76. Strasbourg: Université Louis Pasteur. 143 pp.

Hardie, L. and H. Eugster (1970). "The evolution of closed basin brines." In *Mineralogical Society of America Special Paper* 3, pp. 273–290.

Robert, C. and R. Bousquet (2013). *Géosciences. La dynamique du système Terre.* Paris: Belin. 1159 pp.

Slimani, R., A. Guendouz, F. Trolard, A. S. Moulla, B. Hamdi-Haïssa, and G. Bourrié (2017). "Identification of dominant hydrogeochemical processes for groundwaters in the Algerian Sahara supported by inverse modeling of chemical and isotopic data." In *Hydrology and Earth System Sciences* 21, pp. 1669–1691. DOI: 10.5194/hess-21-1669-2017.

Spencer, R., T. Lowenstein, E. Casas, and P. Zhang (1990). "Origin of potash salts and brines in the Qaidam Basin, China." In. *Fluid-Mineral Interactions: A Tribute to H.P. Eugster.* Ed. by R. Spencer and I.-M. Chou. Special Publication 2. San Antonio, Texas: The Geochemical Society, pp. 395–408.

Tardy, Y. (1969). *Géochimie des altérations - Étude des arènes et des eaux de quelques massifs cristallins d'Europe et d'Afrique.* Mémoires du Service de la Carte géologique d'Alsace-Lorraine 31. Strasbourg: Université Louis Pasteur. 199 pp.

Trauth, N. (1977). *Argiles évaporitiques dans la sédimentation carbonatée continentale et épicontinentale tertiaire : bassins de Paris, de Mormoiron et de Salinelles (France), Jbel Ghassoul (Maroc)*. Sciences Géologiques, Mémoire 49. Strasbourg: Université Louis Pasteur. 195 pp.

List of Authors

Guilhem BOURRIÉ
Académie d'Agriculture
de France (AAF)
Paris
France

Yvette DEWOLF
Paris Diderot University – Paris 7
France

Fernand JOLY (†)
Paris Diderot University – Paris 7
France

Index

Other titles from

ISTE

in

Ecological Science

2020

JOLY Fernand, BOURRIÉ Guilhem
Mankind and Deserts 1: Deserts, Aridity, Exploration and Conquests

BRUSLÉ Jacques, QUIGNARD Jean-Pierre
Fish Behavior 1: Eco-ethology
Fish Behavior 2: Ethophysiology

LE FLOCH Stéphane
Remote Detection and Maritime Pollution: Chemical Spill Studies

ROSSIGNOL Jean-Yves
Climatic Impact of Activities: Methodological Guide for Analysis and Action

2019

AMIARD Jean-Claude
Industrial and Medical Nuclear Accidents: Environmental, Ecological, Health and Socio-economic Consequences
(Radioactive Risk SET – Volume 2)

Nuclear Accidents: Prevention and Management of an Accidental Crisis (Radioactive Risk SET – Volume 3)

BOULEAU Gabrielle
Politicization of Ecological Issues: From Environmental Forms to Environmental Motives

GIRAULT Yves
UNESCO Global Geoparks: Tension Between Territorial Development and Heritage Enhancement

KARA Mohamed Hichem, QUIGNARD Jean-Pierre
Fishes in Lagoons and Estuaries in the Mediterranean 2: Sedentary Fish
Fishes in Lagoons and Estuaries in the Mediterranean 3A: Migratory Fish
Fishes in Lagoons and Estuaries in the Mediterranean 3B: Migratory Fish

OUVRARD Benjamin, STENGER Anne
Incentives and Environmental Policies: From Theory to Empirical Novelties

2018

AMIARD Jean-Claude
Military Nuclear Accidents: Environmental, Ecological, Health and Socio-economic Consequences
(Radioactive Risk SET – Volume 1)

FLIPO Fabrice
The Coming Authoritarian Ecology

GUILLOUX Bleuenn
Marine Genetic Resources, R&D and the Law 1: Complex Objects of Use

KARA Mohamed Hichem, QUIGNARD Jean-Pierre
Fishes in Lagoons and Estuaries in the Mediterranean 1: Diversity, Bioecology and Exploitation

2016

BAGNÈRES Anne-Geneviève, HOSSAERT-MCKEY Martine
Chemical Ecology

2014

DE LARMINAT Philippe
Climate Change: Identification and Projections

Printed in the USA/Agawam, MA
September 16, 2026

910661.017